Livestock Products Technology

NIPA GENX ELECTRONIC RESOURCES & SOLUTIONS P. LTD.
New Delhi-110 034

Livestock Products Technology

V.P. Singh (M.V.Sc., Ph.D)
Officer Incharge
Department of Livestock Products Technology
College of Veterinary & Animal Sciences
Sardar Vallabhbhai Patel University of Agriculture & Technology
Modipuram, Meerut-250110, Uttar Pradesh

NIPA GENX ELECTRONIC RESOURCES & SOLUTIONS P. LTD.
New Delhi-110 034

NIPA GENX ELECTRONIC
RESOURCES & SOLUTIONS P. LTD.

101,103, Vikas Surya Plaza, CU Block
L.S.C.Market, Pitam Pura, New Delhi-110 034
Ph : +91 11 27341616, 27341717, 27341718
E-mail:newindiapublishingagency@gmail.com
www: www.nipabooks.com

For customer assistance, please contact
Phone: + 91-11-27 34 17 17 Fax: + 91-11- 27 34 16 16
E-Mail: feedbacks@nipabooks.com

ISBN: 978-93-90591-28-2

Composed and Designed by NIPA.

सरदार वल्लभभाई पटेल कृषि एवं प्रौद्योगिक विश्वविद्यालय
मेरठ - 250110

Sardar Vallabhbhai Patel University of Agriculture & Technology
Meerut-250110

डо. आर. के. मित्तल
कुलपति

Dr. R. K. Mittal
Vice Chancellor

Phone : 0121-2888522(O)
0121-2888566 (R)
Fax : 0121-2888505(O)
Web : svbpmeerut.ac.in
E-mail : vc2016svpuat@gmail com

No. : SV/PV/C /2020

Date : ....

Foreword

The livestock sector globally is highly dynamic. In developing countries, it is evolving in response to rapidly increasing demand for livestock products. Livestock systems have both positive and negative effects on the natural resource base, public health, social equity and economic growth. Currently, livestock is one of the fastest growing agricultural subsectors in developing countries. Livestock systems occupy about 30 per cent of the planet's ice-free terrestrial surface area and are a significant global asset. The livestock sector is increasingly organized in long market chains that employ at least 1.3 billion people globally and directly support the livelihoods of around 600 million poor smallholder farmers in the developing world. Keeping livestock is an important risk reduction strategy for vulnerable communities, and livestock are important providers of nutrients and traction for growing crops in smallholder systems.

Its share of agricultural GDP is already around 33 per cent and is quickly increasing. This growth is driven by the rapidly increasing demand for livestock products, this demand being driven by population growth, urbanization and increasing incomes in developing countries. At the same time, the expansion of agricultural production needs to take place in a way that allows the less well-off to benefit from increased demand and that moderates its impact on the environment.

The objective of bringing out this Manuscript is to provide the versatile information on various aspects of Livestock products technology to students and field workers in finding out some answers to their problems. 1 am sure that document prepared as "Livestock Products Technology" will serve as a ready reckoner for the subject concerned.

I appreciate the efforts of the author in bringing out this useful publication.

(R. K. Mittal)

Preface

'Livestock Products Technology' is a core course of Veterinary and Animal Science degree programme. At masters level it has its own dignity and students taking admission in this course are getting jobs in various fields. They are also applying in various competitive examinations related to administration, Veterinary Science, Animal Science and Food Science related jobs. The students of Livestock Products Technology course curriculum at Masters and Ph.D level are very well acquainted theoretically and practically on Dairy Technology, Abattoir and By-products Technology, Fresh and Processed Meat Technology, Food Packaging, Food Microbiology, Poultry and Egg Technology, Food Biotechnology and Wool Technology. Most of the competitive examinations in this vital field are mainly based on these subjects. Keeping view of all these aspects a book "Livestock Products Technology" is prepared with most possible multiple choice questions, fill in the blanks, true false and matching type questions. The pattern of questions and most possible answers may be helpful to of Undergraduate, Post Graduate and Ph.D students for various competitive examinations i.e. Junior Research Fellowships, Senior Research Fellowships, National Eligibility Tests, Agricultural Research Services, Civil Services and other competitive examinations.

I am optimistic that the compilation of this book will provide requisite updated knowledge of the subject and will serve the readyreckoner of Livestock Products Technology. The suggestions are heartily invited so that I can improve the manuscript in future.

(V.P.Singh)

Contents

1

Dairy Technology

Q.1 Multiple type questions

1. In UHT treatment, milk is heated at

   a. 65°C for 30 minutes　　b. 71°C for 15 seconds

   c. 10°C for 10seconds　　d. 140°C for 2-3 seconds

2. For making grainy good quality casein, cow milk should be precipitated at ______pH

   a. 4.0　　b. 4.1

   c. 4.6　　d. 5.2

3. For good quality casein, milk should be precipitated at________temperature

   a. 30-32°C　　b. 35-37°C

   c. 40-42°C　　d. 45-47°C

4. BIS standards for fat content in casein is

   a. 1.5%　　b. 2.5%

   c. Nil　　d. No standard

5. _______________will has more minerals

   a. Paneer whey　　b. Cheddar cheese whey

   c. Casein whey　　d. Co-precipitate whey

6. ________________will has least protein content

   a. Paneer whey　　b. Co-precipitate whey

   c. Cheddar cheese whey　　d. Casein whey

7. Ricotta cheese is prepared from

   a. Whey　　b. Whey+milk

   c. Skim milk　　d. Whole milk

8. Maximum possible lactose recovery from cheese
   a. 70% b. 60%
   c. 50% d. 40%
9. The fat content in skim milk for edible casein preparation must be
   a. 0.1% b. <0.06
   c. 0.01% d. Nil
10. Efficient working of casein is essential to
    a. Reduce lactose content b. Reduce ash content
    c. Reduce free acidity d. All of the above
11. Butter milk composition is similar to that of
    a. Skim milk b. Whey
    c. Cow milk d. Buffalo milk
12. Lactose crystallization during whey powder preparation is essential for
    a. More recovery of powder b. Making powder less hydroscopic
    c. Increasing density of powder d. All of the above
13. Hydroscopic properties of lactose are most utilized in
    a. Pharmaceutical tablet preparation b. Seeding of condensed milk
    c. Baby food manufacture d. None of the above
14. Amongst the whey produced in India, ______________is produced in largest quantity
    a. Paneer and chhana whey b. Cheese whey
    c. Casein whey d. Co-precipitate whey
15. ___________has the maximum nutritive value
    a. Casein b. Co-precipitate
    c. Whey proteins d. Cereal protein
16. Amongst the whey produced in the world, __________is produced in largest quantity
    a. Cheese whey b. Paneer and chhana whey
    c. Co-precipitate whey d. Casein whey

17. Which of the statement is correct?
    a. Lactose has more solubility & sweetness than sucrose
    b. Lactose has lower solubility but higher sweetness than sucrose
    c. Lactose has lower solubility and sweetness than sucrose
    d. Lactose has high solubility but lower sweetness than scurose
18. Lactose in whey solids constitutes about
    a. 90% b. 80%
    c. 50% d. 60%
19. Protein content of paneer whey is about
    a. 0.4% b. 0.7%
    c. 0.1% d. 0.2%
20. Quantity of whey produced during casein production is about__________% of skim milk.
    a. Casein b. Co-precipitate
    c. Whey proteins d. Cereal protein
21. Non hygroscopic lactose powder can be prepared by
    a. Controlling the temperature during drying
    b. Proper packaging
    c. Spray drying
    d. Using cheddar cheese whey
22. Name the by-product obtained in the manufacture of cheese
    a. Butter b. Sweet whey
    c. Acid whey d. None of the above
23. The best represented percentage ratios of fat: protein: lactose in cow's milk is
    a. 2.5:3.0:6.5 b 3.7:3.5:4.2
    c. 3.7:3.1:4.2 d. 2.9:3.5:4.1
24. Select any whey protein that has an enzyme function being a coordinate component of lactose synthetase
    a. Serum albumen b. Beta-lactoglobulin
    c. Immunoglobulin d. Alpha lactalbumin

25. The flavoured milk with 1-2% fat level are known as

    a. Skim flavoured milk
    b. Toned flavoured milk
    c. Flavoured drink
    d. Immitation milk

26. Which of the following dairy by-product is claimed to have therapeutic and health promoting properties?

    a. Yoghurt
    b. Acidophilus milk
    c. Kefir
    d. Lassi

27. Milk sugar plays an important role in successful control of the following in condensed milk

    a. Texture
    b. Heat stability
    c. Flavour
    d. Salt balance

28. Which is the correct nitrogen percentage weight on dry basis maximum according to ISI for edible quality casein

    a. 14.0
    b. 14.5
    c. 15.0
    d. 15.5

29. Whey from milk treated with rennet also contains the macropeptide of molecular weight about 8000 that is

    a. Split from K-casein
    b. A subunit of beta lactglobulin
    c. Proteose-peptone fraction
    d. Heat denatured protein of unknown nature

30. Which among the following is not a by-product of Indian dairy industry?

    a. Skim milk
    b. Lassi
    c. Chhana
    d. Ghee residue

31. The difference between plain condensed skim milk and evaporated skim milk is

    a. Former does not contain sugar
    b. Method of manufacture is different
    c. Total solids are more in evaporated skim milk
    d. None of the above

32. The sugar content of flavoured milk generally remains between

    a. 5-7%
    b. 10-20%
    c. 2-5%
    d. None of the above

33. Iron content of good quality khoa is about

 a. 101-103 ppm
 b. 105-108ppm
 c. 110-115ppm
 d. 120-125ppm

34. Stale flavour in khoa is caused by excessively long period of storage at temperature ranging between

 a. 5-10°C
 b. 10-15°C
 c. 15-20°C
 d. 20-25°C

35. The acidity percent of good quality dahi is between

 a. 0.5-0.6
 b. 0.6-0.7
 c. 0.7-0.8
 d. 0.8-0.9

36. One kilogram of fresh sweet milk requires _____gm of citric acid/ lactic acid for coagulation.

 a. 2.0-2.5
 b. 3.0-4.0
 c. 4.0-5.0
 d. 5.0-60.

37. Khoa is_________milk product

 a. Fermented
 b. Partially dehydrated
 c. Dried
 d. Coagulated

38. Acidity in dahi is due to

 a. Acetic acid
 b. Lactic acid
 c. Propionic acid
 d. Formic acid

39. The most common lactose fermenting bacteria is

 a. *Streptococcus lactis*
 b. *Bacillus subtilis*
 c. Coliform
 d. *Bacillus coagulanse*

40. Khoa should contain not less than ___________percent fat

 a. 20
 b. 25
 c. 30
 d. 35

41. Chhana is ___________milk product

 a. Fermented
 b. Fermented and coagulated
 c. Acid and heat coagulated
 d. Rennet coagulated

42. For preparing burfi_______________type khoa is used

a. Pindi
b. Dhap
c. Danedar
d. None of the above

43. Compared to raw milk pressure, normal pressure of pasteurized milk in HTST process is kept at

a. Higher level
b. Lower level
c. Same level
d. None of the above

44. Iso-electric point of casein is at pH

a. 5.4
b. 4.7
c. 4.1
d. 5.0

45. Common soaps are

a. Wetting agents
b. Chelating agents
c. Sequestering agents
d. None of the above

46. As per PFA rules (1976), fat percentage in cow milk in U.P. should not be less than

a. 3
b. 4
c. 3.5
d. 4.5

47. Lactose possess a

a. Beta-2-5-linkage
b. Beta-1-4-linkage
c. Alpha-1-4-linkage
d. None of the above

48. NDDB advocated payment of milk on the basis of its content of

a. Fat only
b. SNF only
c. Fat and SNF only
d. None of the above

49. Vacuum maintained in the third chamber of vacreater is

a. 5 inches Hg
b. 15 inches Hg
c. 26 inches Hg
d. None of the above

50. The time –temperature combination for stassanization process is

a. 74°C for 5 min
b. 74°C for 7 sec
c. 71°C for 15 sec
d. None of the above

51. Teepol is a

a. Wetting agents
b. Chelating agents
c. Sequestering agents
d. None of the above

52. Normal pressure of heating and cooling medium in HTST pasteurizer is maintained at

a. 15 psi
b. 14 psi
c. 12-13 psi
d. 20psi

53. Compared to plate cooler, cooling efficiency of internal tubular cooler is

a. Greater
b. Lower
c. Same
d. None of the above

54. The most common coagulant used in the manufacture of mineral acid casein is

a. Lactic acid
b. Nitric acid
c. Hydrochloric acid
d. Phosphoric acid

55. In the manufacture of low calcium co-precipitate, milk is heated to

a. 72°C for 15 sec
b. 85°C for 15 min
c. 90°C for 1 min
d. 90°C for 15 min

56. Casein curd should be washed with water to remove

a. Impurities
b. Dust particles
c. Excess acid
d. Excess fat

57. The right temperature for drying of casein is

a. 80°C
b. 90°C
c. 60°C
d. 100°C

58. The final cooking temperature for cottage cheese is

a. 37-40°C
b. 42-47°C
c. 49-55°C
d. 56-60°C

59. Cream cottage cheese contains

a. 0.5% fat
b. 4% fat
c. 10% fat
d. 15% fat

60. In cow's milk, alpha-lactalbumen is second in abundance to beta-lactoglobulin. Their relative amounts are approximately

a. 1-2
b. 1-3
c. 1-3.5
d. 1-4

61. In level of diacetyl desired in cultured butter milk is

a. 0.5-1.0ppm
b. 1.5-2.5ppm
c. 2.5-3.5ppm
d. 3.5-4.5ppm

62. Fresh whole milk Ricotta resembles cottage cheese, but is

a. More pasty
b. Moist and grainy
c. More cohesive
d. More firm

63. The pH-viscosity curve for drying of sodium caseinate is at a minimum in the pH range of

a. 6.0-6.5
b. 6.6-7.0
c. 7.1-7.5
d. 7.6-7.9

64. The average specific gravity for cow milk is

a. 1.030
b. 1.036
c. 1.300
d. 1.321

65. Pasteurization and bottling of milk on a large scale for organized distribution was started at Aarey in

a. 1959
b. 1950
c. 1960
d. 1965

66. Contribution of goat to the total milk production in India is about

a. 3%
b. 10%
c. 5%
d. 2%

67. Double toned milk contains

a. 1.5% fat
b. 3 %fat
c. 4.5% fat
d. 6% fat

68. Formalin is a

a. 60% solution of formaldehyde
b. 40% solution of formaldehyde
c. 40% solution of formic acid
d. None of the above

69. Developed acidity in milk is due to

    a. Casein, albumen, citrates, phosphates
    b. Conversion of lactose to lactic acid
    c. Citric acid
    d. None of the above

70. The acidity in mastitic milk is

    a. Higher than normal milk
    b. Equal to normal milk
    c. Lower than normal milk
    d. None of the above

71. The minimum fat% in buffalo milk in Uttar Pradesh, as per PFA rules, 1976 should be

    a. 5.0
    b. 6.5
    c. 6.0
    d. 4.5

72. The percent daily per capita availability of milk in India is

    a. 150g
    b. 144g
    c. 120g
    d. 365g

73. As per PFA standards, the maximum limit for added diacetyl content in desi butter is

    a. 3ppm
    b. 5ppm
    c. 2ppm
    d. Nil

74. The maximum obtainable overrun in creamery butter is

    a. 25%
    b. 30%
    c. 20%
    d. 15%

75. Under Indian conditions, the optimum churning temperature in butter making is

    a. 12-15°C
    b. 5-8°C
    c. 9-11°C
    d. None of the above

76. For long storage, butter is normally stored at

    a. 0°C
    b. -5°C
    c. 12°C
    d. -25°C

77. The permitted antioxidant in ghee is

    a. BHA
    b. BHT
    c. NDGA
    d. Ethyl Gallate

78. The pasteurization process was introduced by the scientist

a. Edward Jenner
b. Louis Pasteur
c. Nicholas Appert
d. Robert Koch

79. The HTST method of milk pasteurization is based on Time-temperature combination of

a. 70°C for 15 seconds
b. 72°C for 15 seconds
c. 70°C for 15 minutes
d. 72°C for 15 minutes

80. The most recently identified pathogen transmitted through contaminated milk is

a. *Brucella melitensis*
b. *Yersinia enterocolitica*
c. *E.coli* O157:H7
d. *Mycobacterium tuberculosis*

81. The only pathogen-specific antimicrobial substance naturally found in the colostrums is

a. Complement
b. Bifidus factor
c. Lactoperoxidase
d. Immunoglobulins

82. ICMSF stands for

a. Indian Council of Microbiological Safety of Foods
b. International Commission on Microbiological Specifications for Foods
c. Indian Commission on Microbiological Specifications for Food
d. None of the above

83. Most of foodborne outbreaks associated with consumption of soft cheese have been reported to be caused

a. *Staphylococcus aureus*
b. *Yersinia enterocolitica*
c. *E.coli*
d. *Listeria monocytogenes*

84. The oxidative or hydrolytic rancidity reactions in milk are catalysed by

a. Oxidase
b. Lipase
c. Phosphatase
d. Catalase

85. In case of mastitis where Listeria organisms are shed in the milk of affected animals, it is recommended

a. To treat the animal
b. To vaccinate the animal
c. To cull the animal
d. To ferment the milk of infected animal

86. A rise 3^{0}C in pasteurization temperature was recommended to ensure the killing of

a. *Staphylococcus aureus* b. *Salmonella typhimurium*
c. *Mycobacterium bovis* d. *Coxiella burnetti*

87. The natural bacteriostatic effect of fresh milk lasts up to

a. 12hrs b. 9hrs
c. 6hrs d. 3hrs

88. Butter milk composition is similar to that of

a. Skim milk b. Whey
c. Cow milk d. Buffalo milk

89. Hydroscopic properties of milk are mostly utilized in

a. Pharmaceutical tablets b. Seeding of condensed milk
c. Baby food manufacture d. Culture media

90. Common alkali used to produce sodium casein is

a. Sodium polyphosphate b. Sodium hydroxide
c. Sodium carbonate d. Sodium by-carbonate

91. The membrane process used in preparation of whey protein concentrate is

a. Reverse osmosis b. Ultrafiltration
c. Microfiltration d. Nanofiltration

92. The maximum obtainable overrun in creamery butter is

a. 30% b. 20%
c. 25% d. 10%

93. Codex Alimentarius Commission was established in the year

a. 1950 b. 1993
c. 1971 d. 1962

94. ISO 22000 relates to

a. Quality management system b. Food safety management
c. Quality assurance system d. Performance Improvement

95. Which of the following dairy product is perceived as functional food
    a. Khoa
    b. Yoghurt
    c. Paneer
    d. Flavoured milk
96. Which of the following is whey coagulated cheese
    a. Cottage cheese
    b. Primost
    c. Sapsago
    d. Gammelost
97. Preservation of evaporated milk is done by
    a. Citrate
    b. Carbonates
    c. Bicarbonates
    d. Benzoates
98. Heart of cheese preparation is
    a. Renneting
    b. Ripening
    c. Cheddaring
    d. Hooping
99. Which of the following principle is most efficiently used in detection of pesticides in dairy products
    a. Infrared spectrometry
    b. Mass spectrometry
    c. Flow injection analysis
    d. Pressure drop
100. Which of the following aromatic volatile is responsible for cooked flavour in dairy products?
    a. Free SH groups
    b. Acetones
    c. Ketones
    d. Lactones
101. Casoxins, a bioactive peptide derived from casein is used as
    a. Anti-stress
    b. Anti-microbial
    c. Immunostimulant
    d. Opioid agonsists
102. Fatty acids characteristics of milk fat is
    a. Linoleic acid
    b. Linolenic acid
    c. Caprylic acid
    d. Butonoic acid
103. Which of the following mineral in milk is not available in ionized from
    a. Sodium
    b. Potassium
    c. Chloride
    d. Calcium

104. Which of the following is not responsible for yhe "weak body" of ice-cream

a. Low total solids
b. High over-run
c. Insufficient stabilizer
d. Too much stabilizer

105. Oxidized flavour development is catalyzed by

a. Vitamin-C
b. Riboflavin
c. Copper
d. Manganese

106. Emulsifier used in ice-cream preparation is

a. CMC
b. Glycerides
c. Glycolates
d. Carageenon

107. The base material for preparation of Srikhand is

a. Khoa
b. Chhana
c. Chakka
d. Panir

108. Under Indian conditions, the optimum churning temperature in butter making is

a. 12-15°C
b. 9-11°C
c. 5-8°C
d. 15-18°C

109. Brownish discolouartion of milk is caused by

a. *Pseudomonas synxantha*
b. *Pseudomonas syncynea*
c. *Pseudomonas putrefaciens*
d. *Serratia marcescens*

110. Bacteria responsible for 'fishy flavour' in milk is

a. *Pseudomonas fragii*
b. *Pseudomonas syncynea*
c. *Aeromonas hydrophila*
d. *Serratia marcescens*

111. The most commonly used radio-isotope for food irradiation is

a. 60Co
b. 137Cs
c. 130I
d. 60Cs

112. Milk is deficient in

a. Iron
b. Calcium
c. Potassium
d. Chloride

113. The share of livestock sector in national GDP is

a. 4.5% b. 2%

c. 3.9% d. 7.1%

114. The anti-microbial properties in modified atmosphere packaging is attributed to

a. CO_2 b. Nitrogen

c. Oxygen d. Argon

115. The milk based pre-biotic derived from lactose is

a. Galactooligosaccharide b. Inulin

c. Fructooligosacchrides d. Glucooligosaccharide

116. Adulteration of milk with water is best judged by determination of

a. Freezing point b. Specific gravity

c. Lactose d. Total solids

117. 12-D concept is mainly concern with

a. *Clostridium botulinum* b. *Bacillus stearothermophilus*

c. *Bacillus cereus* d. *Bacillus sporothermodurans*

118. The test for ensuring adequate sterilization is

a. Phosphatase test b. Turbidity test

c. ATP count d. Resazurin reduction test

119. The aW of intermediate moisture foods range between

a. 0.20-0.40 b. 0.60-0.85

c. 0.10-0.20 d. 0.50-0.60

120. The most variable component of milk is

a. Proteins b. Fat

c. Lactose d. Ash

121. Irradiated milk is produced to increase the content of

a. Vitamin-C b. Vitamin-D

c. Vitamin-A d. Vitamin-A

122. Under PFA rules, minimum fat and SNF content (%) of recombined milk is

a. 3.0 &8.5 b. 3.0& 9.0

c. 4.5 &8.5 d. 4.5 & 9.0

123. The fat content of plastic cream ranges between

a. 65-85% b. 40-50%

c. 25-35% d. 50-60%

124. Whipping quality of cream is best achieved at the fat% of

a. 30-35% b. 40-50%

c. 50-60% d. 60-80%

125. NDDB was established in the year

a. 1946 b. 1970

c. 1965 d. 1982

126. The most commonly used stabilizer of animal origin in ice-cream is

a. Sodium alginate b. Carrageenon

c. Gelatin d. Guar gum

127. The storage temperature of ice-cream should not be less than

a. -5°C b. -12°C

c. -18°C d. -40°C

128. Which of the following is soft and unripened cheese

a. Cheddar b. Spasago

c. Cottage d. Processed cheese

129. Curd tension of soft curd is

a. >25g b. <25g

c. <30g d. >40g

130. Total plate count in 'very good' grade of cow milk should be less than

a. 2.0 million b. 20 million

c. 1.0 million d. 0.2 million

131. Most bacteria, yeast and mold are inhibited at the aW of

a. 0.70
b. 0.75
c. 0.68
d. 0.85

132. The maximum permissible level of migration of substances from the package to the food should not exceed

a. 40ppm
b. 80ppm
c. 100ppm
d. 60ppm

133. The material used for packaging of dairy products must provide barrier to

a. Light
b. Temperature
c. Gases
d. All of the above

134. Sweetening agent used in diabetic ice-cream

a. Invert sugar
b. Corn syrup
c. Sucrose
d. Saccharin

135. The two commonly used frequencies in microwave preservation are

a. 915&2050 MHz
b. 815&2450 MHz
c. 915&2450 MHz
d. 815&2450MHz

136. The best coagulating agent used in preparation of paneer is

a. Acetic acid
b. Alum
c. Vinegar
d. Citric acid

137. Most widely used sanitizer in Indian dairy industry is

a. Iodophores
b. Quaternary ammonium compounds
c. Chlorine
d. Steam

138. Greenish yellow colour of whey is because of

a. Riboflavin
b. Casein
c. Xanthophylls
d. Carotene

139. Pasteurization of milk is first attributed to

a. Aristotle
b. Louis Pasteur
c. Alexander Flemming
d. Dr Soxhlet

140. Which of the following plastic film has highest tensile strength
    a. Polyester b. Nylon
    c. Polypropylene d. HDPE

141. The combined use of two or more preservation techniques is termed as
    a. Hurdle technology b. HACCP
    c. Total quality management d. Shelf stable technology

142. Which of the following dairy product is richest in fat content
    a. Ghee b. Butter
    c. Chhana d. Khoa

143. Lactolin is a natural antimicrobial substance produced by
    a. *Lactobacillus acidophilus* b. *Pediococcus acidilactici*
    c. *Lactobacillus delbrueckii* d. *Lactobacillus plantarum*

144. Which of the following is flexible laminate
    a. Retort pouches b. Tetra-brik system
    c. Liquid pack system d. Conofast system

145. By-product produced during manufacture of casein is
    a. Toned milk b. Whey
    c. Skim milk d. Lassi

146. Chhana is used as base material in manufacture of
    a. Gulabjamun b. Burfi
    c. Sandesh d. Peda

147. Soy paneer is also known as
    a. Surimi b. Tofu
    c. Chakka d. Basundi

148. Most commonly used platform test to determine heat stability and pH of milk
    a. Alcohol test b. Alcohol-alizarin test
    c. Titrable acidity d. COB test

149. Which of the following indigenous sweet is prepared with 'danedar' khoa
    a. Burfi b. Cham-cham
    c. Rasogolla d. Kalakand

150. The texture of condensed milk is primarily affected by

a. Lactose
b. Fats
c. Proteins
d. Ash

151. According to Indian standards, the moisture content(%) of whole milk powder shall not exceed

a. 4
b. 5
c. 6
d. 10

152. Number of grades for ghee specified by 'Agmark' are

a. 3
b. 4
c. 2
d. 1

153. Cheese prepared from goat milk is

a. Cheddar
b. Ricotta
c. Gjetost
d. Primost

154. The fat content (%) in infant milk food range between

a. 18-28
b. 5-15
c. 30-34
d. 35-39

155. The culture used in manufacture of blue cheese is

a. *Penicillium candida*
b. *Penicillium camemberti*
c. *Penicillium claviferne*
d. *Penicillium roqeiforti*

156. The purest grade of lactose is designed as

a. Crude
b. Refined
c. Extra pure
d. USP

157. The loss of ghee is maximum when prepared by

a. Country method
b. Creamer butter method
c. Direct cream method
d. Pre-stratification method

158. According to Indian Standards, the moisture content of edible casein shall not exceed

a. 10
b. 15
c. 5
d. 20

159.Freshly drawn milk should be promptly cooled to

a. 5°C
b. 10°C
c. 12°C
d. 7°C

160. Grade of milk with methylene blue reduction time of 5 hours and above is

a. Very good
b. Good
c. Fair
d. Poor

161. Which of the following indigenous sweet involves use of both khoa and chhana

a. Gulabjamun
b. Pantua
c. Milk cake
d. Sandesh

162. The sugar content of flavoured milk remains between

a. 5-7%
b. 10-13%
c. 2-5%
d. 15-20%

163. The commonly used heating temperature of skim milk for preparation of co-precipitate is

a. 72°C
b. 80°C
c. 90°C
d. 95°C

164. The commonly used acid in manufacture of casein is

a. Lactic
b. Hydrochloric
c. Sulfuric
d. Phosphoric

165. Which of the following is lactic acid –alcohol fermented milk

a. Kumiss
b. Bulgarian milk
c. Acidophilus milk
d. Cultured buttermilk

166. The rancid flvour in ice-cream is caused due to

a. Fat hydrolysis by lipase
b. Over heating of mix
c. Inadequate sugar
d. Low quality ingrédients

167. Direct Microscopic Count (DMC) of 'good' grade milk should be

a. <0.5 million
b. <5 million
c. <1.5 million
d. <2 million

168. Adulteration of ghee with vegetable fat is detected by

a. Baudouin test
b. Iodine test
c. Hehners test
d. Leech test

169. Common soaps are

a. Chelating agents
b. Sequestering agents
c. Wetting agents
d. Sanitizing agents

170. Normal pressure of heating and cooling medium in HTST pasteurizer is maintained at

a. 5-8 psi
b. 8-10psi
c. 12-13psi
d. 16-18 psi

171. Which of the following will have least protein content

a. Paneer whey
b. Casein whey
c. Co-precipitate
d. Cheese whey

172. Non-hygroscopic lactose powder can be prepared by

a. Controlled drying process
b. Proper packaging
c. Spray drying
d. Cheddar cheese whey

173. In UHT treatment milk is heated at

a. 65°C/30 min
b. 71°C/15 sec
c. 140°C/2-3 sec
d. 100°C/10sec

174. Maximum tolerance limit (in ppm) of 2-4D in milk is

a. 0.05
b. 0.75
c. 0.50
d. 0.10

175. Which of the following milk vitamin in special concentration provide protection against sunlight

a. Niacin
b. Thiamine
c. Pantothenic acid
d. Riboflavin

176. The ageing temperature for ice-cream should not exceed

(a) 6.0°C
(b) 10°C
(c) 4.5°C
(d) 15°C

177. Fat recovery in indigenous method of ghee making is

(a) 80-85% (b) 90-92%

(c) 85-90% (d) 92-95%

178. Which among is not a process of continuous butter making

(a) Fritz process (b) Alfa level process

(c) Cherry Burrell process (d) Hennery Johnson process

179. Overrun in butter is due to

(a) Moisture (b) Curd

(c) Salt (d) All of the above

180. Khoa is a

(a) Partially dehydrated whole milk product

(b) Dried milk product

(c) Coagulated whole milk product

(d) Evaporated toned milk product

181. Best quality Channa can be obtained from

(a) Buffalo milk (b) Cow milk

(c) Goat milk (d) Mixed milk

182. Which among is the soft cheese

(a) Cottage (b) Cheddar

(c) Emmental (d) Parmesan

183. Cheddar cheese was originated in

(a) England (b) Switzerland

(c) Canada (d) Ireland

184. Edible casein is obtained from

(a) Skim milk (b) Whey

(c) Both of the above (d) None of the above

185. Whey is rich in

(a) Lactose (b) Casein

(c) Both of the above (d) None of the above

186. Metallic flavor in ghee is generally observed due to

(a) Fat oxidation (b) Smoky fire

(c) Contact of copper and iron (d) Incorrect straining

187. Which among is not a type of khoa

(a) Pindi (b) Dhap

(c) Danedar (d) Nirmal

188. In this packaging system, package is the integral part of food

(a) Active packaging (b) Intelligent packaging

(c) Edible packaging (d) Smart packaging

189. Crumbly body in cheese is due to

(a) Excess fat (b) Excess salt

(c) Excess moisture (d) None of the above

190. Overrun in ice-cream may be maximum upto

(a) 50% (b) 100%

(c) 80% (d) 20%

191. ....milk is preferred for preparation of high quality khoa

(a) Cow (b) Buffalo

(c) Goat (d) None of the above

192. Which among is not the theory of butter churning

(a) Fisher & Hooker (b) Kings Modern

(c) Rah'n Foam (d) Stan's gel

193. Yeasty flavor in butter is due to

(a) Use of old cream (b) Hydrolysis by lipase

(c) Over pasteurization (d) None of the above

194. Greasy appearance is the characteristics of

(a) Cow milk Channa (b) Buffalo milk Channa

(c) Both of the above (d) None of the above

195. Which of the following is true for table butter

(a) Not less than 80 percent fat (b) Not more than 3 percent salt

(c) Both of the above (d) None of the above

196. Casein/fat ratio in cheese milk should be

(a) 0.58-0.62 (b) 0.62-0.65

(c) 0.68-0.70 (d) 0.72-0.75

197.Iron content of khoa prepared from cow milk is

(a) 100 (b) 101

(c) 102 (d) 103

198. Neutralizer best suited to ice-cream is

(a) Sodium sulfate (b) Sodium bicarbonate

(c) Sodium silicate (d) None of the above

199. Yield of cow milk paneer is

(a) 16-18% (b) 18-20%

(c) 20-22% (d) >22%

200. Which of the following is a acid cum heat coagulated milk product

(a) Paneer (b) Khoa

(c) Basundi (d) None of the above

201. Which of the following is not a food product prepared from whey

(a) Ricotta cheese (b) Dried whey

(c) Lactose (d) None of the above

202. Pilot sterilization test is used for

(a) Evaporated milk (b) Fermented milk

(c) Pasteurized milk (d) None of the above

203. Cheddaring consists of

(a) Packing (b) Turning

(c) Pilling (d) All of the above

204. Kumiss contains

(a) 1.0% alcohol (b) 2.5% alcohol

(c) 1.0% lactic acid (d) 1.0% alcohol & 2.5% lactic acid

205. Foamy meltdown in ice-cream is due to_______________during freezing

(a) Incorporation of too much air (b) Incorporation of less air

(c) Use of excess stabilizer (d) None of the above

206. Bactofugation removes .... Percent bacteria by

(a) 98 (b) 99
(c) 90 (d) 95

207. The titrable acidity of cow milk is

(a) 0.11-0.12 (b) 0.12-0.13
(c) 0.13-0.14 (d) 0.14-0.15

208. Which of the following shows greatest daily variation in milk composition?

(a) Protein (b) Sugar
(c) Fat (d) Ash

209. Which of the following is not a dairy Sanitizer?

(a) Steam (b) Hot water
(c) Iodophor (d) None of the above

210. Which of the following is not true for sterilized milk

(a) 50 percent of vitamin C is destroyed
(b) Has distinct rich flavour
(c) Phosphatase test is used for assessing completeness of process
(d) Forms soft digestible curd

211. Yield of Surati paneer from cow milk is

(a) 24 (b) 25.5
(c) 28.5 (d) 32

212. Fish eye is a common defect encountered in.........production

(a) Ice cream (b) Butter
(c) Cheese (d) Dahi

213. Starter in cheese is added @

(a) 0.5-1.0 percent (b) 1.0-1.5 percent
(c) 1.5-2.0 percent (d) 2.0-2.5 percent

214. Overrun in ice cream is due to

(a) Air incorporation (b) Freezing
(c) Fat (d) None of the above

215. For preparation of condensed milk, the concentration of milk solids is

(a) 1:2 (b) 1:2.5
(c) 1:3 (d) 1:3.5

216. Seeding is related to which of the following

(a) Cheese (b) Condensed milk
(c) Both of the above (d) None of the above

217. Milk powder should contain not less than...... percent fat

(a) 24 (b) 25
(c) 26 (d) 28

218. Which of the following is not true for soft curd milk

(a) Has low casein content (b) Low calcium content
(c) Has curd tension 25g (d) None of the above

219. Which of the following is not a coagulated milk product

(a) Paneer (b) Shrikhand
(c) Rabri (d) Dahi

220. Which of the following is/are skim milk based product/s

(a) Casein (b) Quarg
(c) Cottage cheese (d) All of the above

221..............milk is preferred for preparation of high quality khoa

(a) Cow (b) Buffalo
(c) Goat (d) None of the above

222. Irradiated milk has increased content of

(a) Vitamin-A (b) Vitamin-B
(c) Vitamin-C (d) Vitamin-D

223. Toned milk contains

(a) 3.0% fat and 8.5%SNF (b) 1.5 % fat and 8.5%SNF
(c) 3.0% fat and 9.0%SNF (d) 4.5 % fat and 8.5%SNF

224.Test for assessing sterilized milk

(a) Turbidity test (b) Chlorine test
(c) Phosphatase test (d) None of the above

225.Which of the following is true for table butter

(a) Not less than 80 percent fat
(b) Not more than 3 percent salt
(c) Both of the above
(d) None of the above

226. Casein/fat ratio in cheese milk should be

(a) 0.58-0.62
(b) 0.62-0.65
(c) 0.68-0.70
(d) 0.72-0.75

227. Iron content of khoa prepared from cow milk is

(a) 100
(b) 101
(c) 102
(d) 103

228. Neutralizer best suited to ice-cream is

(a) Sodium sulfate
(b) Sodium bicarbonate
(c) Sodium silicate
(d) None of the above

229. Energy value of buffalo milk is

(a) 75C/100g
(b) 85C/100g
(c) 95C/100g
(d) 100C/100g

230. Which of the following is a acid cum heat coagulated milk product

(a) Paneer
(b) Khoa
(c) Basundi
(d) None of the above

231. Which of the following is not a food product prepared from whey

(a) Ricotta cheese
(b) Dried whey
(c) Lactose
(d) None of the above

232. Pilot sterilization test is used for

(a) Evaporated milk
(b) Fermented milk
(c) Pasteurized milk
(d) None of the above

233. Cheddaring consists of

(a) Packing
(b) Turning
(c) Pilling
(d) All of the above

234. Kumiss contains

(a) 1.0% alcohol
(b) 2.5% alcohol
(c) 1.0% lactic acid
(d) 1.0% alcohol & 2.5% lactic acid

235. Which of the following is true for sterilized milk

(a) Lack of cream layer formation (b) 90% of Vitamin C is lost

(c) Accurate Gerber test (d) Does not require refrigerated storage

236. Water requirement for dairy processing plant should be

(a) Soft water (b) Hardness 112 mg/litre

(c) Both of the above (d) None of the above

237. COB test is performed in milk for the assessment of

(a) Acidity (b) Mastitic milk

(c) Colostrums milk (d) All of the above

238. Microbial rennet can be prepared from

(a) Yeast (b) Mold

(c) Both of the above (d) None of the above

239. Animal rennet can be prepared from

(a) Calf stomach (b) Pig stomach

(c) Goat stomach (d) All of the above

240. Yoghurt is a fermented dairy product, produced by the action of

(a) *Streptococcus salivarius* (b) *Lactobacillus delbrueckii*

(c) Both of the above (d) None of the above

241.Traditionally used stabilizers for the commercial manufacturing of ice-cream

(a) Gelatin (b) Guar Gum

(c) Both of the above (d) None of the above

242. Traditionally stabilizers for the commercial manufacture of ice-cream are now being replaced by ________________ for better quality

(a) Low-viscosity gum acacia (b) Sodium alginate

(c) Modified carob bean gum (d) All of the above

243. Pasteurization of butter improves it qualities like

(a) Flavour (b) Keeping quality

(c) Both of the above (d) All of the above

244. Neutralization once used in cream then they reduces loss of

(a) Butter fat (b) Control abnormal flavours

(c) Increases keeping quality (d) All of the above

245. Indicate the factor affecting the churning

(a) Fat richness (b) Acidity

(c) Viscosity (c) All of the above

246. Name the theory of churning

(a) Phase inversion (b) Foam theory

(c) Both of the above (d) None of the above

247. Yoghurt is produced by symbiotic action between__________in a ratio of 1:1

(a) *Lactobacillus bulgaricus* (b) *Streptococcus thermophillus*

(c) Both of the above (d) None of the above

248. The antibacterial activity of raw milk is contributed by natural inhibitors like

(a) LP system (b) Lysozyme

(c) Lactoferrin (d) All of the above

249. Aseptic packaging requires that a sterilized product be packed into

(a) Sterilized package (b) Sterilized surrounding

(c) Both of the above (d) None of the above

250. The carton packages for UHT milk are composed of polyethylene film containing

(a) Aluminium foil (b) Paper

(c) Polyethylene (d) All of the above

251. The legal quality requirement of standardized milk is

(a) Fat 4.5% (b) SNF 8.5%

(c) Both of the above (d) None of the above

252. Trisodium phosphate in a detergent helps in

(a) Dissolution of proteins (b) Fat Emulsification

(c) Both of the above (d) None of the above

253. The kefir grains contains

(a) *Streptococcus lactis*
(b) Beta bacterium caucasicum kefir bacilli
(c) Lactose fermenting yeast
(d) All of the above

254. Name the atomizer used for spray drying of milk

(a) Nozzle atomizers
(b) Rotary atomizer
(c) Both of the above
(d) None of the above

255. Name the heaters generally used in indirect heating of milk in UHT processing

(a) Plate exchangers
(b) Tubular exchangers
(c) Scraped-surface exchangers
(d) All of the above

256. In milk, casein stability is extremely sensitive to

(a) Hydrophobicity
(b) pH
(c) Both of the above
(d) None of the above

257. Indicate the true statement for milk constituents

(a) Fat is in emulsion form
(b) Protein and some mineral matters in colloidal suspension
(c) Lactose together with some minerals and soluble proteins in true solution
(d) All of the above

258. For cleaning of aluminium alloy dairy utensils, detergent used should be

(a) Strong alkali
(b) Weak alkali
(c) Strong acid
(d) None of the above

259. Q10 value provides a quick check of how sensitive reaction rate is to the

(a) Temperature
(b) Time
(c) Pressure
(d) None of the above

260. Heat required to evaporate 1 kg water with inlet air temperature 215°C is

(a) 5000 kg in single stage drier
(b) 6000 kg in single stage drier
(c) 7000 kg in single stage drier
(d) None of the above

261. High WPN-index is observed in ___________ heat milk powder.

(a) Low
(b) Medium
(c) High
(d) None of the above

262. Solubility index is mainly a measure of coagulated

(a) Casein (b) Whey protein

(c) Both of the above (d) None of the above

263. Spray drying is usually concentrated to ____________ percent solids.

(a) 40 (b) 50

(c) 60 (d) None of the above

264. Immediate cooling of milk powders helps in reducing

(a) Browning (a) Free fat

(a) Sulphydryl compounds (a) None of the above

265. Oxidized or metallic flavour in milk may develop due to

(a) Direct contact of milk with metal like copper

(b) Exposure of milk to iron

(c) Both of the above

(d) None of the above

266. A clarifixator is a machine developed in Sweden functions as

(a) Clarifiers (b) Homogenizer

(c) Both of the above (d) None of the above

267. Indicate the true option related to the manufacture of Paneer from buffalo milk

(a) First heating of milk to 80°C (b) Cooling of milk to 70°C

(c) Coagulant addition in cooled milk (d) All of the above

268. Alcohol-Alizarin test is used in milk to determine

(a) Stability (b) pH

(c) Both of the above (d) None of the above

269. As per PFA rules the standardized milk for liquid consumption should contain a minimum of______________ throughout the country

(a) 4.5% Fat (b) 8.5 %SNF

(c) Both of the above (d) None of the above

270. Milk salts in milk are present in_____________ state.

(a) Soluble (b) Colloidal

(c) Both soluble and colloidal (d) None of the above

271. Indicate the true statement for AGMARK standards of ghee

(a) Special (red lebel) RM value -28, Max FFA 1.4

(b) General (green label), RM value -28, Max FFA 2.5.

(c) Both of the above

(d) None of the above

272. Indicate the true statement for milk homogenization

(a) Milk temperature 60-65°C (b) Pressure 2000-2500 lb psi

(c) Both of the above (d) All of the above

273. Indicate the lactic acid –alcohol fermented milk product

(a) Kefir (b) Kumiss

(c) Both of the above (d) All of the above

274. The maximum moisture and acidity contents in ghee as per legal requirements is

(a) 0.5% & 3% acidity as oleic acid (b) 0.3% & 5% acidity as oleic acid

(c) 0.7% & 3% acidity as oleic acid (d) All of the above

275. Total production of milk is converted into khoa is

(a) 7% (b) 5%

(c) 3% (d) 10%

276. Total production of milk is converted into Chhana is

(a) 5% (b) 4%

(c) 2% (d) 6%

277. As per the latest PFA rules, khoa shall not have less than

(a) Moisture 20% and not < 70% fat

(b) Moisture 10% and not < 80% fat

(c) Moisture 15% and not < 75% fat

(d) None of the above

278. Chhana is used as a base material for preparation of

(a) Rasogolla (b) Sandesh

(c) Both of the above (d) None of the above

279. Indicate the true statement for PFA act on ghee

(a) Moisture not >0.5% (b) FFA not >3%

(c) Both of the above (d) None of the above

280. Foam spray dried milk has

(a) Less bulk density and improved dispersability

(b) Less bulk density and reduced dispersability

(c) More bulk density and improved dispersability

(d) More bulk density and reduced dispersability

281. Water activity' is reduced in evaporation process due to

(a) Protein denaturation (b) Fat oxidation

(c) Increase in concentration of solids (d) None of the above

282. In an instantizer, agglomeration results in ___of interstitial air between the particles

(a) Increase (b) Decrease

(c) Stable (d) None of the above

283. In standardization of milk for condensed milk, ratio of Fat : SNF is usually

(a) 1:2.44 (b) 1:2.52

(c) 1:2.63 (d) 1:2.70

284. Name the compound promotes development of oxidized flavour in ghee during storage at room temperature.

(a) Oxygen (b) Copper

(c) Iron (d) All of the above

285. Important processing steps in developing desirable texture in khoa is

(a) Stirring and Scrapping (b) Heating and standardization

(c) Preheating and acid treatment (d) None of the above

286. Fat content in butter and ghee respectively is

(a) 80 and 99.5% (b). 75 and 99.5%

(c) 80 and 99% (d) None of the above

287. Kheer contains

(a) 8% fat and 9% sugar (b) 9% fat and 8% sugar

(c) 10% fat and 5% sugar (d) None of the above

288. Shelf life of kheer can be increased by

(a) Pasteurization and cooling
(b) Heat sterilization and addition of Nisin
(c) Both of the above
(d) None of the above

289. Dahi is a rich source of

(a) Vit-B
(b) Vit-C
(c) Both of the above
(d) None of the above

290. Culture used in preparation of sweet dahi is

(a) *Streptococcus lactis*
(b) *Streptococcus diacetilactis*
(c) *Streptococcus cremoris*
(d) All of the above

Q.2 Fill in the blanks with suitable words:

1. The pH of freshly drawn cow milk ranges between________________.
2. Moisture contents in ghee should not be more than________________.
3. The diacetyl content in butter should not exceed________________.
4. ____________ acidity in milk is due to casein, citrates, phosphates and dissolved Co_2.
5. BIS lactometer is calibrated at a temperature of________________.
6. Milk samples for chemical analysis are preserved with________________.
7. The specific acidity of fat is________________.
8. The capacity of milk pipette used for fat estimation is________________
9. The minimum SNF content in double toned milk is________________.
10. The pH of freshly drawn buffalo milk ranges between ________________.
11. Moisture contents in dehydrated milk products should not be more than______________.
12. Natural colour added in butter is ________________.
13. Titrable acidity in normal milk should not be beyond ________________.
14. Alcohol test is performed in milk for________________.
15. Scale on butyrometer ranges from ________________.
16. Phenolphthalein used in estimation of titrable acidity gives __________ colour in alkalis.
17. The specific acidity of cow milk is________________.

18. The minimum fat content in toned milk is ________________.
19. Flavour production in butter is due to ______________________.
20. Butter is a ________________ type of emulsion.
21. Average iodine value of butter is _______________.
22. Acidophillus milk has _____________ acid flavour.
23. Organism used in Bulgarian milk production is ________________.
24. Good quality yoghurt has _________ lactic acid.
25. Ropiness in raw milk is caused by growth of ________________ at low temperature.
26. *Psudomonas putrifaciens* bacteria cause surface taint in butter at ____________.
27. Black discolouration of butter is caused by______________________.
28. Aerobic spore former ______________________ causes gassiness in cheese.
29. The most common spoilage pattern displayed by bacteria in cottage cheese is______________.
30. Bitterness is commonly observed with ____________cheese cultures.
31. Butter must be protected from loss of _____________ which causes change in colour.
32. In commercial manufacture of casein skim milk is pasteurized at ________________.
33. Average moisture content of 30 mesh size casein will be _____________ percent.
34. In manufacture of acid casein recommended strength of mineral acid is_____________.
35. Rennet casein is precipitated at _____________ pH.
36. Higher pH of precipitation leads to casein with__________ fat content.
37. Large lumps of casein dry_________________.
38. Counter flow system of washing casein curd require ____________ amount of water than conventional method.
39. Heating of caseinate solution causes the viscosity to ________________.
40. Yield of commercial casein from 1000kg skim milk will be __________.

41. Common alkali used to produce sodium casein is_________________.
42. Commonly used acid in manufacture of casein is _________________.
43. The ratio of wash water to skim milk generally varies form ___________.
44. The commonly used drier for drying casein is _______________.
45. The membrane process used in preparation of whey protein concentrate is__________.
46. The commonly used heating temperature of skim milk for preparation of co-precipitate is_____________.
47. A bactofuge is a centrifuge that separates the ___________ from milk.
48. Homogenized milk becomes tastier than unhomogenized milk because of__________________.
49. The water activity (a_w) is expressed by _____________________.
50. 'Best-before-date' labeling denotes_________________________.
51. Toning of buffalo milk is a sort of ________________standradization.
52. The sequence of wash cycles involves water rinse, alkali wash, water rinse and acid wash and final water rinse are the steps of____________________.
53. Skimming efficiency of a centrifugal cream separator is determined by______________________.
54. ____________ is principally vacuum treatment of milk or cream to remove off flavour.
55. Gross chilling efficiency of a farm bulk cooler is calculated as __________________.
56. Packaging materials are best sterilized before use by_______________.
57. ________________ is a centrifuge for separation of dirt and other sediment material.
58. Bulk vending system of milk distribution has become popular in the Indian cities because _____________________.
59. _________________ performs three functions viz. clarify, standardize and separate.
60. Cold milk separator has lower capacity than a conventional separator of comparable size because of_____________________.
61. Long duration cooling of raw milk would eliminate the problem of mesophiles but may lead to bitterness due to ____________________.

62. Caramelized flavour in milk is due to reaction between _______________.
63. Coconut flavour in milk is originated due to chemical reaction between _____________________.
64. The isoelectric point of casein is___________________.
65. Ghee residue is a good source of___________________________.
66. Primost cheese contains approximately ________________ fat.
67. Casein exists in milk as a____.
68. The size of the lactose crystals which can cause sandiness defects is_________________.
69. Solids from heated whey are precipitated by the addition of______________ for manufacture of whey cheeses.
70. The acidity of sweet dahi should be maximum % weight lactic acid as ______________.
71. According to PFA (1976) the standardized milk for liquid consumption should contain a minimum of _______________ solid not fat.
72. The key aroma compound in dahi is ____________________.
73. The main compound responsible for yogurt aroma is _______________ compound.
74. For obtaining quality of rashogolla the chhana prepared from________milk is recommended.
75. For obtaining quality of peda, the khoa prepared from _______________ milk is used.
76. While making chhana the coagulation temperature of milk should be about___________.
77. The shelf life of khoa is _____________ at 37±1°C.
78. For making good quality khoa, milk is standardized to ____________ fat.
79. Continuous khoa making equipment was developed at ______________.
80. Among the value added indigenous dairy products, maximum profit is expected from______________.
81. Keeping in view the legal standards of khoa it's expected out turn-out of 100 litres of milk with 15% total solid would range within _____________________.
82. Chhana refers to the milk solids obtained by________________coagulation of boiled hot whole milk and subsequent drainage of whey.

83. Triprocess machine performs separation, clarification and ____________.
84. MBR time for Grade-I milk should not be less than ______________.
85. Flow diversion valve is an important unit of____________________.
86. In 18:8 stainless steel, iron content is ____________________.
87. The active alkalinity of can washing solution should be ____________.
88. Normally chlorine solution containing ______________ available chlorine is used for sanitizing dairy equipments.
89. Lactose possesses a potentially free ______________ group.
90. The seeding material used in manufacture of lactose is ____________.
91. The membrane system used for concentration of whey is called as ______________.
92. Lactose can be hydrolyzed with the help of ____________________.
93. The number of washings recommended for the preparation of acid casein is __________.
94. The dairy industry would prefer to manufacture ____________ from skim milk rather than sodium caseinate because of health considerations.
95. Cheese whey contains ____________________.
96. The amount of heat coagulable protein in casein whey is ___________.
97. The heat stability of milk ____________ progressively with increasing concentration of milk solids.
98. The separation of solids materials in the bottom of the container is known as ____________.
99. The major uses of casein are made in foods, adhesives, manmade fibres, plastics and ____________.
100. Sterilization of foods in hermetically sealed containers using ionizing radiation is called________________.
101. The most commonly used radio isotope for food irradiation is________.
102. The safety limit for energy of betaparticles to ensure freedom from induced radio activity in irradiated foods is________________.
103. Gamma irradiation at a dose not exceeding ______________ has been approved by FAO/WHO for all foods.
104. The linear accelerator is a good source of _______________ for food irradiations.

105. The most commonly used microwave frequencies are _______________ MHz.

106. The device converting electrical energy into microwave radiation is called__________.

107. A maximum of ___________ of microwave leakage from a new microwave oven is considered safe for health.

108. Localized heating near the surface of frozen food during thawing by microwave is called_________________.

109. _____________ CIP system can be more cost effective than conventional CIP system for large plants.

110. Polyphosphates are useful as________________agents in dairy detergents.

111. Enzyme based detergents are used for cleaning of____________membranes.

112. ________________ can be used to prevent corrosion of tin by cleaning acids or alkali.

113. __________________is a good example of detergent –cum-sanitizer.

114. An absorbed radiation dose equal to 0.01J/Kg of the absorbing medium is called_____________.

115. The expanded form of rep is ______________________.

116. Te penetration power of a microwave radiation is determined by its _______________ which bears a direct relationship with the depth of penetration.

117. ___________________cleaning system saves on time of cleaning but requires extra alkali.

118. Sugar ratio in SCM is maintained at about_____________________.

119. The most common storage defect in dried skim milk powder is _________________.

120. Pilot sterilization test is used for_____________.

121. The commonly used seed material for sweetened condensed milk is_______________.

122. The time temperature combination generally used for sterilization of evaporated milk is ___________________________.

123. The optimum total solids required for spray drying is about___________.

124. The moisture content in skim milk powder according to the BIS should not be more than__________________.

125. Complete sterilization of milk is tested by__________________test.

126. Wettability term describes the ability of powder particles to__________ on the surface of the particles.

127. Cationic surfactants are poor detergents but excellent __________________________.

128. In an evaporator, basic heat transfer is______________________________ _________.

129. For manufacture of WMP, SNF: Fat ratio should be __________________.

130. Level of undenatured whey protein in milk powders is indicated by __________________________.

131. Solubility index of milk powders is the ________________________.

132. Bactofugation process helps in removal of________________________.

133. Coenzyme __________________ is the substance that is naturally present in the human body, with the highest levels in the heart, liver, kidneys, and pancreas.

134. Z-value of spore is __________________________.

135. Z-value of vegetative organisms is ____________________.

136. Average __________________ is determined by the average velocity of the particles in the product and dimensions of holding tube.

137. Preheating of milk to temperature ____________________results in better shelf life of powder.

138. Solubility percentage of partly skim milk powder (sour) should not be less than _______________ percent.

139. Sanitization implies the destruction of almost all ____________________ organisms.

140. Pasteurization of milk ____________________ the cream line.

141. Negative phosphatase test indicate__________________ pasteurization.

142. Alcohol (Ethanol) test is used to determine __________________ of milk.

143. HTST pasteurizing equipment occupies ____________space than LTLT pasteurizing equipment.

144. Rancid flaour in milk is due to fat hydrolysis caused ________________ enzyme.

145. Acid used in estimation of milk by Gerber method is ________________ acid.

146. PFA requirement for standardized milk is____________ percent fat.

147. Buffalo milk contains ________________ protein than cow milk.

148. Skim milk weight ____________ than cow milk.

149. Time temperature combination used in HTST pasteurizer is ____________.

150. Time temperature combinations are used in batch type sterilizer is____________.

151. The various tests performed at the reception dock are called____________.

152. Homogenization of milk is done to obtain ______________________.

153. Buffalo milk has ______________ heat stability than cow milk.

154. According to PFA, fat content in paneer should be ________________.

155. The normal fat content in efficiently separated skim milk is______________.

156. The normal level of inoculums used in Dahi preparation during winter months generally is ____________________.

157. A continuous method of ghee manufacture has been developed at NDRI using the principle ___________________________.

158. According to PFA requirements the desi butter should contain ____________fat.

159. The normal overrun in butter is_________________.

160. The normal concentration of acid solution used for coagulating milk for chhana manufacture is __________________.

161. Bulk raw milk, when received from a chilling centre in the dairy factory shall not have a temperature more than __________________.

162. Clarifier removes the suspended foreign particles by _______________.

163. Under efficient operating conditions of the separator, the fat content of the skim milk is in the range of ____________________.

164. Standardization of milk refers to the ______________________.

165. Lactose can be manufactured from _________________.

166. For market milk purposes, the homogenization pressure under in 2nd stage is________.

167. Pasteurization refers to___________________________.

212. Optimum temperature of incubating milk for dahi preparation is__________.

213. Rabri is generally prepared by condensing milk at simmering temperature followed by blending with ______________@7% of milk.

214. _______________% alcohol is used for performing alcohol precipitation test.

215. Time-temperature of vat-pasteurization was increased from 61.7°C to 62.8°C to ensure the destruction of the organism named__________________.

216. The natural bacteriostatic effect of fresh milk lasts up to ____________ hours after milking.

217. Milk is deficient in vitamin____________.

218. Pasteurized milk should not have SPC more than________________.

219. Phosphatase test depends on release of free___________by enzymatic reaction.

220. Ropiness of milk is caused by ____________________microbe.

221. Milk is deficient in the much needed mineral called_______________.

222. The process of removing micro-organism from milk by centrifugation is known as_________________.

223. ________________ refers to the process by which dried milk and milk products are made soluble.

224. The artificial sugar produced from lactose is called____________________.

225. Ghee residue is good source of ___________________.

226. Flow diversion valve is important unit of _____________________.

227. Normally chlorine solution containing _______________available chlorine is used for sanitizing dairy equipments.

228. The membrane system used for concentration of whey is _______________.

229. The heat stability of milk _____________ progressively with increasing concentration of milk solids.

230. The flouring agent used in butter is ___________________.

231. Sterilization of foods in hermetically sealed containers using ionizing radiation is called______________________.

232. Enzyme based detergents are used for cleaning of _____________membranes.

233. Pilot sterilization test is used for ________________________.

234. The optimal total solids content required for spray drying is______________.

235. Polyphosphates are useful as _________________ in dairy detergents.

236. Cryovac is one of the method of________________________.

237. ____________ standards deal with manufacturer having their own product research and development.

238. _____________ organism are indicator of faecal contamination in milk and milk products.

239. RITARD model of testing is done to identify food poisoning due to __________________.

240. Chinese dairy industry was recently affected adversely due to contamination of dairy products with_____________________.

241. MMPO was established in the year ____________________.

242. The milk fat content of ice-cream should not be less than _____________.

243. _______________test is used to evaluate all plastic material intended for food contact application.

244. _________________ films are most commonly used in vacuum packaging.

245. The specific gravity of ghee is_____________________.

246. ______________________ test is employed to check the efficiency of pasteurization.

247. The average over-run in softy is_____________________.

248. Gouda cheese is preferably prepared from___________ milk.

249. APEDA stands for____.

250. Best before date labeling denotes__________________________________.

251. In modified atmosphere packaging, nitrogen is used as________________ gas.

252. Surface taint with apple odour in butter is caused due to _________________.

253. Sodium sulphite is used as_____________ along with detergents to prevent corrosion of utensils.

254. Sterilized milk shall remain fit for human consumption for atleast ________ days at room temperature.

255. Putrefaction of food is spoilage under_______________ conditions.

256. The time necessary for killing given number of organisms at specific temperature is known as____________________.

257. Surface ropiness in milk is casued by________________________.

258. The mould ___________________is associated with formation of buttons on the surface of sweetened condensed milk.

259. Average iodine value of butter is______________.

260. __________________ organism is used for production of invertase.

261. According to PFA, the fat content of milk toffee should not be less than _________ on dry matter basis.

262. Maximum tolerance limit (according to PFA) for chlorpyrifos in milk is __________.

263. _______________ organism present in milk produces DNase, a heat stable substance used its detection.

264. The iso electric point of casein is at pH __________________.

265. The cane sugar content of sweetened condensed milk should not be less than _________.

266. According to BIS, fat% in whole milk powder should not be less than _______.

267. The linear accelerator is a good source of_____________ for food irradiation.

268. The main rationale of drying is to___________________water activity.

269. Ordinary whole milk powder contains_____________________fat in solid.

270. Pure clarified fat derived from milk or desi butter or cream is termed as__________.

271. ____________________test is used to detect adulteration of ghee with vanaspathi.

272. Working of butter is also called___________________.

273. Paneer is______________________type of milk product.

274. Misty Dahi is traditionally produced in ______________________.

275. ______________are used to prevent the formation of objectionable large ice-crystals in ice-cream.

276. Standardization of milk is done by using__________________________ method.

277. Ideal temperature for renneting in cheese making is ________________.

278. Kumiss is ________________ fermented milk.

279. ______________ is a dairy product having characteristics intended to cultured milk and soft cheese.

280. Major whey protein is____________________.

281. Kefir is a________________________beverage fermented with kefir grains.

282. Whey protein concentrate is obtained by________________process.

283. As per PFA standards dried milk powder contains not more than______ moisture.

284. Hardening and storage of ice cream is done at ___________°C.

285. As per PFA recommendation hard cheese should contain_________ percent milk fat.

286. As per PFA standards butter contains minimum of__________percent fat.

287. RM value of ghee is not normally less than_______________.

288. The typical yellow colour of butter is due to ___________________.

289. Overrun in ice-cream is mainly due to _______________________.

290. Course texture of ice-cream is mainly due to the formation of _________ during freezing.

291. Dahi showing typical characteristics like brown colour, cooked caramelized flavor is called__________________.

292. Acidity of sweet dahi is ___________________.

293. _______________milk is preferred for making paneer.

294. Ricotta cheese is made from ________________whey.

295. Pearson's square method is used to _______________the milk.

296. The total acidity expressed as lactic acid should not exceed______________ percent in skim milk powder.

297. Bitter flavor in butter is due to_____________________________.

298. Recently GCMMF launched _______________milk claiming better health/ diabetes management.

299. Milk production in India (2019-20) is _________________million tones.

300. ___________________test is used to detect inadequate pasteurization.

301. High percentage of ________________ causes sandiness in ice cream.

302. ________________ is a test for assessing sterilized milk.

303. The ________________ is the heart of cheese.

304. Micro-organisms responsible for lactic fermentation of yogurt are ________________.

305. The casein/fat ratio in cheese milk is standardized to ________________.

306. Homogenization causes sub-division of original fat globules to ________________ (diameter).

307. Reichert-Meissl value of ghee should normally be less than ________________.

308. Acid-rinse in a CIP system utilizes acid solution of ________________ percent acidity.

309. Gerber test is generally performed to estimate ________________ in milk.

310. As per PFA recommendation hard cheese should contain ________________ percent milk fat.

311. Pasteurized milk should have a SPC/ml(or g) ________________.

312. Daily per capita consumption of milk in India is ________________.

313. Acid-rinse in a CIP system utilizes acid solution of ________________ percent acidity.

314. Commonly chlorine solution with ________________ available chlorine are used as sanitizers in dairies.

315. Time-temperature combination ________________ is used for UHT.

316. ________________ is the index organism (bacteria) for pasteurization.

317. Isoelectric point of casein is ________________.

318. The total acidity expressed as lactic acid should not exceed ________________ percent in skim milk powder.

319. Optimum cream churning temperature during butter making under Indian conditions is ________________.

320. Rennet is added after confirming the desired rate of ________________ in cheddar cheese preparation.

Q.3 State whether following statements are True or False

1.	Yield of paneer from cow milk is around 20%.	( )
2.	Whey is yellowish in colour due to riboflavin.	( )
3.	Buffalo milk is preferred for channa making.	( )
4.	Milk fat in butter should not be less than 80%.	( )
5.	Specific gravity of milk is increased on skimming.	( )
6.	Methyline blue reduction test is a direct microscopic count method.	( )
7.	Richmond's scale is not suitable for milk temperature above 25^{0}C.	( )
8.	As per PFA standards, fat contents in buffalo milk in U.P. should not be less than 6.0%.	( )
9.	The 80^{0}C is an optimum temperature for adding sodium citrate in paueer preparation.	( )
10.	Casein is not a protein of whey.	( )
11.	Yield of channa from buffalo milk is 20%.	( )
12.	SPC/ml of pasteurized milk should not exceed 30,000.	( )
13.	Cow milk is preferred for paneer making.	( )
14.	Moisture content in butter should not be more than 20%.	( )
15.	Specific gravity of milk is increased on addition of cream.	( )
16.	Dye reduction tests are rapid methods for determining bacteriological quality of milk.	( )
17.	Specific gravity of milk is inversely related to the temperature.	( )
18.	As per PFA standards, fat contents in cow milk in U.P. should not be less than 4.0%.	( )
19.	LTLT method of pasteurization is preferred over HTST method.	( )
20.	Natural acidity of milk is directly related to its SNF content.	( )
21.	Whey is a concentrated solution of protein.	( )
22.	Casein is a heterogeneous mixture of several individual casein components.	( )
23.	Maximum precipitation of casein occurs around pH4.6.	( )
24.	Casein micelle stability is destroyed by the action of proteolytic enzyme such as pepsin.	( )
25.	Cheddar cheese whey has lower pH than paneer whey.	( )
26.	Common form of lactose is crystalline alpha monohydrate.	( )
27.	Action of heat on skim milk helps to induce interaction between casein and whey protein.	( )
28.	After precipitation casein should be held in whey for sometime in order to produce good casein.	( )
29.	Cooking of casein curd can be done by direct steam injection.	( )
30.	Low acidity during precipitation of casein cause rubbery curd.	( )

31.	Cooking of casein curd is one to improve its solubility.	( )
32.	Tempering process causes moisture in dried casein to equalize.	( )
33.	Sodium caseinate has high affinity for water.	( )
34.	Finally ground casein has more flowability than spray dried sodium caseinate.	( )
35.	Consumption of wash water in counterflow system will be more than conventional washing of casein.	( )
36.	Low calcium co-precipitates have more calcium than rennet casein.	( )
37.	Lactic acid casein is generally made by direct addition of lactic acid to skim milk.	( )
38.	The pH of wash water should be the same as that of casein curd.	( )
39.	Rennet casein and high calcium co-precipitates have similar calcium contents.	( )
40.	Pasteurization of skim milk improves the colour of casein.	( )
41.	Acid casein is easily soluble in water.	( )
42.	Co-precipitates are produced by the action of alkali on casein.	( )
43.	Lactose is less soluble than glucose.	( )
44.	Ultrafiltration will remove only water molecules from whey.	( )
45.	For better grinding temperature dried casein should be kept high.	( )
46.	A creaming cylinder with an inner diameter of 3.6 cm and a sample volume of 250 ml can be used as a measuring device for proof of homogenization.	( )
47.	In the regeneration section of HTST pasteurizer heat is exchanged between milk and hot water.	( )
48.	LP system of preservation is equally effective for poor quality milk.	( )
49.	Milk is homogenized at 60-75^{0}C because the viscosity increases and thus less pressure is required.	( )
50.	Tactile property of milk gives a sensation of graininess of chalkiness to the mouth.	( )
51.	The flow diversion valve is a two way valve designed to control the milk flow in HTST pasteurizers.	( )
52.	Deposits on heat exchange surfaces in dairy plants are largely influenced by the whey proteins and not salts.	( )
53.	The concentration of dissolved calcium and phosphate in milk is decreased by heat treatment.	( )
54.	In long time pasteurization, heated milk is held at 62-65^{0}C either by means of a pipe of appropriate length or by plates.	( )
55.	The plate coolers are efficient in heat transfer because they are operated on direct expansion refrigeration system.	( )

56.	Higher the SNF in the mix greater is the acidity of ice-cream.	( )
57.	Higher the fat content lower is the pressure required for homogenization.	( )
58.	Stabilizers bind water and prevent formation of large ice-crystals.	( )
59.	Emulsifiers bind water and prevent formation of large ice-crystals.	( )
60.	The freezing point of the ice-cream mix is not affected by the stabilizers.	( )
61.	Hydrolysis of lactose prevent sandiness die to emulsification.	( )
62.	Cheese whey has higher protein content than chhana whey.	( )
63.	Acid whey is richer in calcium and phosphate than sweet whey.	( )
64.	Acid casein and rennet casein are structurally similar.	( )
65.	Whey protein concentrate are poor emulsifying agents.	( )
66.	Lassi is the by-product obtained from churning of ripened cream.	( )
67.	Ghee residue produced from desi butter has the least protein content.	( )
68.	Gammelost cheese is a semi soft blue mold cheese.	( )
69.	Alpha-lactose hydrate is the non crystalline form of lactose.	( )
70.	Irradiated milk essentially contains vit-D.	( )
71.	Hydrolysis of lactose increases the sweetness of the product.	( )
72.	Sterilized milk does not need refrigerated storage.	( )
73.	Sodium caseinate has better keeping quality than isoelectric casein.	( )
74.	With increase of pH of precipitation the calcium concentration of casein also increases.	( )
75.	Both primost and Mysost cheese are made from goats' milk.	( )
76.	Baker's yeast is unsuitable for the fermentation of whey.	( )
77.	The whey obtained from cheddar cheese making is more acidic than paneer whey.	( )
78.	The concentration whey proteins in paneer whey is higher than cheese whey.	( )
79.	The stability of casein micelle is disturbed by the action of proteolytic enzymes such as pepsin.	( )
80.	Heat induces interaction between casein micelle and whey proteins.	( )
81.	The casein after precipitation should be kept under whey for some time to produce good quality casein.	( )
82.	Cooking of casein curd can be done by the direct steam injection.	( )
83.	Cooking of casein curd is done to improve its solubility.	( )
84.	Lactic acid casein is generally made by direct addition of lactic acid to skim milk.	( )

85.	The pH of wash water should be the same as that of casein curd.	( )
86.	Rennet casein has higher calcium than acid casein.	( )
87.	The casein content is higher in co-precipitate than acid casein.	( )
88.	The solubility of acid casein is high in water.	( )
89.	The ash content is less in acid casein than rennet casein.	( )
90.	The nutritive value of casein is higher than whey proteins.	( )
91.	Lactose is sweeter than sucrose.	( )
92.	Cow milk gives more yield of casein than buffalo milk.	( )
93.	Lactose is responsible for sandiness in ice-cream.	( )
94.	The TS of butter milk is similar to that of skim milk.	( )
95.	One molecule of lactose gives three molecules of lactic acid.	( )
96.	Lower moisture content of khoa will result in higher keeping quality and vice-versa.	( )
97.	Fat content of buffalo milk khoa is higher than that of cow milk khoa.	( )
98.	Lactose content of chhana is higher than that of milk.	( )
99.	The fat content of chhana is higher when prepared from cow milk than that from buffalo milk.	( )
100.	In order to obtain a desirable body and texture of chhana the pH of coagulation should be around 5.4.	( )
101.	A satisfactory strength of coagulating acid solution is 1-2%.	( )
102.	Chhana from buffalo has a slightly hard body and greasy and coarse texture.	( )
103.	A minimum fat level of 5.0% in cow milk is necessary for production of desirable body and texture in chhana for sweet making.	( )
104.	Delayed straining causes the chhana to retain more moisture than immediate straining.	( )
105.	The flavour of chhana should be mildly acid.	( )
106.	Sanitization with chlorine solution is not recommended for tinned milk cans.	( )
107.	Bottling of milk is more economical than packaging in polyethylene pouches.	( )
108.	Flow diversion valve is operated on air pressure.	( )
109.	Pasteurization reduces the cream line of milk.	( )
110.	Viscolizer machine operates at a lower pressure than homogenizer.	( )
111.	Quality of milk after cold clarification is poorer than that after warm clarification.	( )
112.	In LTLT method of pasteurization, milk is heated to 60^0C for 30 minutes.	( )

113.	Pasteurization destroys useful lactic acid bacteria in milk.	( )
114.	Late lactation milk may cause bitter flavour.	( )
115.	In single stage homogenizer, 2000-2500 psi pressure is applied for milk containing 6% fat.	( )
116.	The maximum precipitation of casein occurs at pH 5.3.	( )
117.	An equally high quality acid casein can be produced by the use of sulphuric acid.	( )
118.	In the production of Ricotta cheese, whey obtained during the manufacture of other cheeses is used as the raw material.	( )
119.	Gjetost is dark brown in colour and possesses a coarse texture.	( )
120.	Mysost is made from cow's skim milk.	( )
121.	The presence of other substances in the lactose solution will not interfere with the molecular movement and orientation to retard crystallization.	( )
122.	Rennet whey from cheddar cheese conatins lesser protein than acid whey from casein manufacture.	( )
123.	Time temperature combinations in the production of sterilized flavoured milk are determined on the basis of inactivation of microorganisms rather than on sensory changes.	( )
124.	Preheating of skim milk to a temperature sufficient to denature whey proteins and to form co-precipitates with casein may not help in increased yield of cottage cheese.	( )
125.	Ghee residue is a good source of methionine.	( )
126.	Alpha rays are most suitable radiation for food preservation.	( )
127.	Gamma rays are more penetrative as compared to high energy beta particles.	( )
128.	Cobalt -60 has a half life of 5.27 years.	( )
129.	The principal hurdle in application ionizing radiation for the preservation of milk products is the burnt flavour produced by irradiation.	( )
130.	The Indian food laws permit preservation of dairy products by means of ionizing radiation.	( )
131.	Gamma irradiation can effectively prevent sprouting in potatoes but not in onion.	( )
132.	Rapid surface heating of food products is the main advantage of microwave processing.	( )
133.	Food held in metal cans cannot be heated by microwaves.	( )
134.	Of all body parts eyes are more sensitive to microwave radiation.	( )
135.	Presence of water in a food is necessary for it to be heated in a microwave field.	( )

136.	Bio detergents are the detergents used for removing biological fouling.	( )
137.	Surfactants are effective in wetting of the soil during cleaning of dairy equipments.	( )
138.	Sodium hydroxide is normally not used in alkali cleaning of aluminium cans.	( )
139.	Acid cleaning intended to remove milk stones from heat exchanger surface is usually carried out at 85^{0}C.	( )
140.	Chemical sterilization is the preferred method for sanitization of the UHT plant.	( )
141.	Increase in ionic calcium in milk leads to improving the heat stability.	( )
142.	Evaporated milk is more popular in our country than sweetened condensed milk.	( )
143.	The particle size of a dried product has no effect on the separation efficiency of a cyclone.	( )
144.	The flowability of dried milk can be improved by addition of sodium aluminium slilcate.	( )
145.	The viscosity of sweetened condensed milk increases logarithmically with storage temperature and arithmetically with storage time.	( )
146.	The optimum temperature for forced crystallization is 50^{0}C.	( )
147.	Homogenization is not necessary in the manufacture of evaporated milk.	( )
148.	The fat content in skim milk powder according to PFA should not be more than 5%.	( )
149.	Forewarming is done to improve the heat stability of milk for the manufacture of condensed milk.	( )
150.	The density of sweetened condensed milk is measured with lactometer.	( )
151.	Latent heat of vaporization increases with the increase in pressure.	( )
152.	Surfactants are stable under alkaline conditions but unstable under acidic conditions.	( )
153.	In single effect evaporation 1.2 kg steam is required to evaporate 1 kg water from milk.	( )
154.	Pilot sterilization test in evaporated milk processing helps in determining the stability of evaporated milk for sterilization.	( )
155.	Microbial bloating in condensed milk is due to contamination and growth of yeast *Torula lactic condensii*.	( )
156.	In roller drying, the product is in contact with the roller for 10 seconds at a temperature of 200^{0}C.	( )

157.	In sterilization 12 decimal reductions for *Clostridium botulinum* is required.	( )
158.	Lysine loss is more in UHT processed milk than in sterilized (glass) milk.	( )
159.	UHT milk shows greater tendency to thicken and coagulate during storage than sterilized milk.	( )
160.	The heat to which milk is subjected to the various manufacturing stages contribute to the resultant WPN index.	( )
161.	Sulphydryl compounds are released on heating of milk and milk products and act as antioxidants.	( )
162.	Powders produced by high pressure nozzle atomizer contain less entrapped air.	( )
163.	In direct method of UHT process, microbial inactivation is greater while chemical damage is less.	( )
164.	Biological value of UHT processed and in bottle sterilized milk is same.	( )
165.	Solubility of milk powder depends upon stability of casein.	( )
166.	Addition of water to milk results in the lowering of its lactometer reading.	( )
167.	Casein exists only in milk and is found in the form of calcium-caseinate-phosphate complex.	( )
168.	Sterilization of dairy equipments refers to complete destruction of all microorganisms from equipment surface.	( )
169.	Water is a minor constituent of milk .	( )
170.	Desi method of ghee making leads to large size crystal (granular) formation in ghee.	( )
171.	The function of Float Controlled Balance Tank (FCBT) in HTST pasteurizer is to maintain a constant head of milk for feeding the raw milk pump.	( )
172.	Milk freezes at temperature higher than water due to the presence of lactose.	( )
173.	The sediment test will show visible foreign matter contained in milk.	( )
174.	Detergent mixture used for cleaning aluminium cans should not be more than 5.0 percent alkalinity.	( )
175.	Properly pasteurized Grade-I milk will have MBR time of over $5^{1}/_{2}$ hours.	( )
176.	Milk sugar is called lactose.	( )
177.	Milk normally contains higher percentage of casein than milk sugar.	( )
178.	One ml of (N/10) sodium hydroxide solution will neutralize 0.009 gm lactic acid.	( )

179.	Pasteurization destroys all bacteria.	( )
180.	Cooling milk below 5⁰C helps in increasing its keeping quality.	( )
181.	Bactofugation is a process for removing about 99.0% bacteria in milk.	( )
182.	A substance that either alone or in a mixture reduces the work required of a cleaning process is known as detergent.	( )
183.	Freezing point of milk is higher than freezing point of water because of soluble substances.	( )
184.	It is possible to get 80.0% regeneration efficiency in HTST pasteurizer.	( )
185.	Copper and its alloy are not preferred in the manufacture of dairy equipments because they taint milk.	( )
186.	The clot-on-boiling (COB) test determines the acidity of milk.	( )
187.	The objects of collection cum chilling centers are to preserve the quality of raw milk supply and to provide easily transport to processing dairy.	( )
188.	Stanssanization is a method of pasteurization of milk by plate type heat exchangers.	( )
189.	Gravity type bottles fillers fill more rapidly than vacuum type milk fillers.	( )
190.	Flow diversion valve is used for holding milk for certain period.	( )
191.	LTLT method of pasteurization is suitable for large scale handling of milk.	( )
192.	The material for construction of a road milk tanker is mild steel.	( )
193.	The object of pasteurization of milk is to kill all the pathogenic and spoilage organisms.	( )
194.	The COB test is used to determine the status of salt balance in milk.	( )
195.	Alkaline phosphatase requires more drastic heat treatment for its destruction than *Mycobacterium tuberculosis*.	( )
196.	The amyl alcohol is used in Gerber test for preventing charring of milk.	( )
197.	The developed acidity in milk is due to the presence of caseinates and phosphates.	( )
198.	The optimum pH for coagulation of milk during paneer manufacture is 4.7.	( )
199.	The optimum temperature for coagulation of cow milk for chhana manufacture is 82⁰C.	( )
200.	Continuous method of butter manufacture results in better control on the composition of resulting butter.	( )
201.	The granulation is better in cow milk ghee as compared to buffalo milk ghee.	( )

202.	Curdy ghee is preferred in South India.	( )
203.	Buffalo milk is preferred for preparation of Rasogolla.	( )
204.	Hydrolytic rancidity is a major cause of ghee spoilage during storage.	( )
205.	Freshly drawn milk is amphoteric in litmus.	( )
206.	The pH of normal and fresh cow milk usually varies from 6.8-7.0.	( )
207.	Milk was first transported in refrigerated chamber through Indian railways from Anand to Bombay in 1955.	( )
208.	In nature , lactose is only found in milk and meat.	( )
209.	The minimum standards prescribed by the PFA (1976) rules for cow milk in Uttar Pradesh is 3.0% fat and 8.5% SNF.	( )
210.	Payment according to fat content of milk encourages adulteration with water.	( )
211.	In milk pricing policy ensures a common approach to cow and buffalo milks.	( )
212.	Collection of milk by cooperative organizations eliminates profit making middlemen.	( )
213.	The acidity of milk is equal to natural acidity plus developed acidity.	( )
214.	Low sediment content in milk does not necessarily mean a low bacterial count in milk.	( )
215.	Dilution of water to milk increases the lactometer reading.	( )
216.	Road transport of milk is cheaper than rail transport over short distance.	( )
217.	Chilling of milk by surface water has the risk of air borne contamination?	( )
218.	Pasteurized milk should have a SPC/ml count not more than 30,000.	( )
219.	Uperization of milk kills all pathogenic organisms.	( )
220.	Common milk micro-organisms grow best between 15-20^{0}C.	( )
221.	Freezing point of milk is not affected by pasteurization.	( )
222.	Addition of formalin in milk interferes with the fat test.	( )
223.	Elements of milk by volume is not affected by temperature.	( )
224.	Boiling test determines the heat stability of milk.	( )
225.	Casein exists only in milk.	( )
226.	In India, total production of cow milk is higher than buffalo milk.	( )
227.	Hydrolysis of lactose yields glucose and fructose.	( )
228.	Higher the SNF content in milk, higher is its natural acidity.	( )
229.	Ghee and butter oil contain clarified fat.	( )
230.	Skim milk is heavier than whole milk.	( )

231.	Pasteurized milk shows positive phosphatase test.	( )
232.	Butter and cheese starter cultures consist of same bacteria.	( )
233.	The pasteurization temperature of milk and ice cream mix are different.	( )
234.	Ripening of cream is essential in butter making.	( )
235.	It is necessary to neutralize the sour cream before manufacturing butter.	( )
236.	Homogenization of milk increases the surface area of fat globules in milk.	( )
237.	The proteins are not denatured in Khoa making.	( )
238.	Sour milk is preferred for preparation of chhana as it coagulates faster.	( )
239.	Addition of water raises the freezing point of milk.	( )
240.	The % of lactose in dahi is lower than that of milk from which it is prepared.	( )
241.	Homogenized milk is more viscous than un-homogenized milk.	( )
242.	Neutralization of sour cream prevents fat losses in butter milk.	( )
243.	Salted butter has shorter storage life than sweet cream butter under commercial cold storage.	( )
244.	Ripened cheese develops characteristics flavours than the green cheese.	( )
245.	Immediate cooling of ghee to refrigeration temperature improves its grainularity.	( )
246.	The amount of residue obtained during ghee preparation increases with the increase in % fat of cream.	( )
247.	Preparation of ghee from cream is preferred over butter method because of more recovery of fat in the former case.	( )
248.	Cow milk is preferred for khoa preparation.	( )
249.	Yeast and molds are primarily responsible for spoilage of ghee during storage.	( )
250.	Whole milk is preferred over skim milk for large scale manufacturing of srikhand.	( )
251.	Srikhand normally contains 45% sugar.	( )
252.	Cow milk chhana is softer than buffalo chhana because of higher moisture in the former.	( )
253.	Slightly sour milk produce better textured kalakand than the fresh milk.	( )
254.	Casein and whey proteins are not significantly affected during manufacturing of khoa.	( )
255.	Buffalo milk containing 14% milk solids will yield more khoa than cow milk having same level of milk solids.	( )

256.	All the whey proteins are transferred into whey during manufacture of chhana.	( )
257.	The main Bacteria used in dahi preparation are *Streptococcus lactis* and *Streptococcus cremoris.*	( )
258.	Sugam Dairy at Baroda manufactures srikhand and gulabjamun at large scale	( )
259.	Delhi is biggest marketing centre for khoa.	( )
260.	Use of buffalo milk for manufacturing of dahi gives better texture than cow milk.	( )
261.	Strip cup test is used for detecting the mastitic milk	( )
262.	In case the bacterial count of milk is high, methylene blue takes more time to discolourize.	( )
263.	Presence of nitrate indicates milk adulteration with water.	( )
264.	Iodine test is used for detection of starch in milk.	( )
265.	Phosphatase may get reactivated in UHT pasteurized milk.	( )
266.	Clot-on-boiling test gives sufficiently advance warning of impending souring.	( )
267.	The temperature of milk should not exceed 10^{0}C till it reaches processing unit.	( )
268.	The titrable acidity gives idea about the buffering capacity of milk.	( )
269.	In case, a higher concentration of alcohol is used, the sensitivity of alcohol test is decreased.	( )
270.	The resazurin test will never reject the milk with good shelf life.	( )

Q.4 Match the following

A.

A		B		
1.	Protein concentration in butter	a.	26.4	( )
2.	Protein concentration in dried whole milk	b.	7.0	( )
3.	Protein concentration in evaporated milk	c.	0.6	( )
4.	Protein concentration in cottage cheese	d.	12.9	( )
5.	Protein concentration in dry whey	e.	17.0	( )

6.	Protein concentration in market milk	f.	3.5	( )
7.	Gammelost	g.	*Melilotus coerulea*	( )
8.	Sapsago	h.	Goat milk whey	( )
9.	Gretost	i.	Fermented skim milk	( )
10.	Quarg	j.	*Pencillium roqueforti*	( )

B.

A		B		
1.	Phosphatase test	a.	Bromothymol blue	( )
2.	Antibiotic residues	b.	Dye réduction test	( )
3.	Titrable acidity	c.	Sédimentation test	( )
4.	Détergent test	d.	Milk acidity	( )
5.	Muddy water adultération	e.	Chromatography	( )
6.	Glucose detection	f.	Phosphomolybdic acid reagent	( )
7.	Bacterial count	g.	Bromocresol purple	( )
8.	Urea detection	h.	Par-nitro phenyl phosphate	( )
9.	Mastitis milk	i.	Incipient souring of milk	( )
10.	Alizarin-alcohol test	j.	Alkaline pH	( )

Key

Dairy Technology

Q.No. 1							
1.d	41.c	81.d	121.b	161.b	201.d	241.c	281.c
2.b	42.a	82.b	122.a	162.a	202.a	242.d	282.a
3.b	43.a	83.d	123.a	163.c	203.d	243.c	283.a
4.a	44.a	84.b	124.a	164.b	204.b	244.d	284.d
5.c	45.a	85.c	125.c	165.a	205.a	245.d	285.a
6.b	46.c	86.d	126.c	166.a	206.b	246.c	286.a
7.b	47.b	87.d	127.b	167.a	207.c	247.c	287.a
8.a	48.c	88.a	128.c	168.a	208.b	248.d	288.c
9.b	49.c	89.a	129.b	169.c	209.d	249.c	289.c
10.d	50.b	90.b	130.d	170.c	210.c	250.d	290.d
11.a	51.a	91.b	131.d	171.c	211.c	251.c	
12.b	52.c	92.c	132.d	172.a	212.c	252.c	
13.a	53.b	93.d	133.d	173.c	213.a	253.d	
14.a	54.c	94.b	134.d	174.a	214.a	254.c	
15.c	55.d	95.b	135.c	175.d	215.b	255.d	
16.a	56.c	96.b	136.d	176.c	216.b	256.c	
17.c	57.c	97.a	137.c	177.a	217.c	257.d	
18.b	58.c	98.b	138.a	178.d	218.d	258.b	
19.b	59.b	99.b	139.d	179.d	219.c	259.a	
20.b	60.b	100.a	140.a	180.a	220.d	260.a	
21.a	61.b	101.d	141.a	181.a	221.b	261.a	
22.b	62.b	102.d	142.a	182.a	222.d	262.a	
23.b	63.b	103.d	143.d	183.a	223.a	263.b	
24.d	64.a	104.d	144.a	184.a	224.a	264.b	
25.c	65.b	105.c	145.d	185.a	225.c	265.c	
26.b	66.a	106.c	146.c	186.c	226.c	266.c	
27.a	67.a	107.c	147.b	187.c	227.d	267.d	
28.b	68.b	108.b	148.b	188.c	228.b	268.c	
29.a	69.b	109.c	149.d	189.b	229.d	269.c	
30.c	70.c	110.c	150.a	190.b	230.a	270.c	
31.d	71.c	111.a	151.a	191.b	231.d	271.c	
32.a	72.d	112.a	152.c	192.d	232.a	272.c	
33.a	73.d	113.a	153.c	193.c	233.d	273.c	
34.a	74.a	114.a	154.a	194.b	234.b	274.a	
35.b	75.c	115.a	155.d	195.c	235.d	275.a	
36.a	76.d	116.a	156.d	196.c	236.c	276.b	
37.b	77.a	117.a	157.c	197.d	237.d	277.a	
38.b	78.b	118.d	158.a	198.b	238.c	278.c	
39.a	79.b	119.b	159.a	199.a	239.a	279.c	
40.a	80.c	120.b	160.a	200.a	240.c	280.a	

Q.No. 2			
1.	6.4-6.6	41.	Sodium hydroxide
2.	0.3-0.5	42.	HCl
3.	0.2ppm	43.	1:1
4.	Natural	44.	Tray
5.	27^{0}C	45.	Ultrafiltration
6.	Potassium dichromate	46.	90^{0}C
7.	3.2	47.	Micro-organisms
8.	10.75ml	48.	Increased surface area of fat globules
9.	9.0	49.	Amount of available water in foods
10.	6.6-6.8	50.	Date of minimum durability
11.	0.5%	51.	Downward
12.	Annato	52.	CIP System
13.	0.21	53.	% total fat recovered in cream.
14.	Milk acidity	54.	Vacreation
15.	1-10	55.	BTUextracted per hr/ BTU rated per hr x100
16.	Faint pink colour	56.	Gamma radiation
17.	0.10 to 0.26% lactic acid	57.	Milk clarifier
18.	3.0	58.	Convenience and quality assurance
19.	Diacetyl and acetoin	59.	Triprocess machine
20.	Water in oil	60.	Lower milk flow rate
21.	32	61.	Growth of psychrotrops
22.	Strong	62.	Lactose and proteins
23.	*Lactobacillus bulgaricus*	63.	Lactones from milk fat
24.	0.9%	64.	4.6-4.7
25.	*Alcaligenes viscolactis*	65.	Phospholipids
26.	4-7^{0}C	66.	30%
27.	*Pseudomonas nigrifaciens*	67.	Calcium caseinate-calcium phosphate complex
28.	*Bacillus polymyxa*	68.	15-20μ
29.	Slimy curd	69.	Citric acid
30.	Cheddar	70.	0.70%
31.	Moisture	71.	8.5%
32.	72^{0}C	72.	Diacetyl
33.	11	73.	Carboxyl
34.	0.5N	74.	Cow
35.	6.6	75.	Buffalo
36.	Lower	76.	82^{0}C
37.	Slower	77.	07 days
38.	Less	78.	4-5%
39.	Decrease	79.	NDRI
40.	30	80.	Dahi

81.	21-23 Kg	117.	O.R. or over-ride
82.	Acid	118.	62.5-64.0%
83.	Standardization	119.	Oxidized
84.	$5^1/_2$ hrs	120.	Evaporated milk
85.	HTST pasteurizer	121.	Lactose hydrate
86.	74%	122.	116-118^{0}C for 15 minutes
87.	0.5-0.6%	123.	40-50%
88.	150-200ppm	124.	4%
89.	Aldehyde	125.	Microbiological
90.	Lactose	126.	Be wetted by liquid
91.	Reverse osmosis	127.	Excellent biodegradability
92.	Beta galactosidase	128.	By the convection and conduction equations
93.	One	129.	2:1
94.	Potassium caseinate	130.	Whey protein nitrogen index (WPNI)
95.	7% total solids	131.	Ability of a powder to dissolve in water
96.	0.6%	132.	Microorganisms from milk using centrifugal force
97.	Decrease	133.	Q10
98.	Sedimentation	134.	4.38 to 5.44
99.	Paper coating	135.	5.70 to 14.31
100.	Radappertization	136.	Residence time
101.	Cobalt-60	137.	90^{0}C
102.	10 MeV	138.	75%
103.	1Mrad	139.	Pathogenic
104.	Electrons or beta-particles	140.	Decreases
105.	2450 and 915 MHz	141.	Complete
106.	Magnetron	142.	Heat stability
107.	1 mw/cm^2	143.	Less
108.	Run away heating	144.	Lipase
109.	Reuse and multiuse	145.	Sulphuric
110.	Chelating (water softening)	146.	4.5
111.	Cellulosic	147.	More
112.	Sodium sulphite	148.	Less
113.	Jodophor	149.	72^{0}C for 15 seconds
114.	Rad	150.	121^{0}C for 15min
115.	Roentgen equivalent	151.	Plateform test
116.	Wavelength	152.	Equal distribution of fat in milk

153.	Less	187.	Glucose
154.	50%	187.	Glucose
155.	0.01-0.05%	187.	Glucose
156.	1-2%	187.	Glucose
157.	Scrapped surface heat exchanger	187.	Glucose
158.	75% fat	187.	Glucose
159.	22-25%	187.	Glucose
160.	1%	187.	Glucose
158.	75% fat	188.	Colloidal
159.	22-25%	189.	0.7 %
160.	1%	190.	0.2%
161.	7^0C	191.	28%
162.	Centrifugal force	192.	Chakka
163.	0.5%	193.	21%
164.	Adjustment of fat and /or SNF	194.	18%
165.	Milk	195.	Over 20%
166.	500psi	196.	80-100%
167.	Heating and cooling of milk	197.	Lypolysis
168.	Steam	198.	Autooxidation
169.	45 marks	199.	85-90%
170.	Copper	200.	95-97
171.	Pouches	201.	115-117^{0}C
172.	03	202.	80-85^{0}C
173.	15	203.	Hard or grainy
174.	6.7	204.	-SH
175.	Cream	205.	5%
176.	Lighter	206.	65%
177.	3	207.	Hot and acid
178.	More	208.	115^{0}C
179.	Chromium	209.	Lacquered
180.	Less	210.	Bakery
181.	Skim milk	211.	0.75%
182.	Lighter	212.	22^{0}C
183.	Oil in water	213.	Sugar
184.	Larger	214.	68-70%
185.	7.8	215.	*Coxiella burnetti*
186.	Lower	216.	At least one hour

217.	Vit-C	257.	*Alkaligenes viscolactis*
218.	30,000	258.	*Aspergilus nigor*
221.	Iron	259.	32
222.	Bactofugation	260.	*Pullularia pullulens*
223.	Instantization	261.	4%
224.	Lactulose	262.	0.01ppm
225.	Phospholipids	263.	*Staphylococcus aureus*
226.	HTST pasteurizer	264.	4.7
227.	150-200ppm	265.	40%
228.	Reverse osmosis	266.	26%
229.	Decreases	267.	Electrons
230.	Diacetyl	268.	Reduce
231.	Radappertization	269.	25-28%
232.	Cellulosic	270.	Ghee
233.	Evaporated milk	271.	Baudman
234.	45-50	272.	Kneading of butter
235.	Chelating agents	273.	Heat cum acid coagulated
236.	Vacuum packaging	274.	West Bengal
237.	ISO-9001	275.	Stabilizers
238.	Coliform	276.	Pearson square method
239.	*Compylobacter jejuni*	277.	30°C
240.	Melamine	278.	Lactic acid
241.	1992	279.	Yoghurt
242.	10%	280.	Beta Lactoglobulin
243.	Global migration	281.	Self carbonated
244.	Polypropylene	282.	Ultrafiltrations
245.	0.93-0.94	283.	5%
246.	Phosphatase	284.	-23 to -29°C
247.	30-50%	285.	42%
248.	Cow	286.	80%
249.	Agriculture and Processed Food Export Development Authority	287.	28
250.	Date of minimum durability	288.	Annatto
251.	Inert	289.	Incorporation of air
252.	*Chrysobacterium josteii*	290.	Large ice crystals
253.	Inhibitor	291.	Misty dahi
254.	7	292.	0.7
255.	Anaerobic	293.	Buffalo milk
256.	TDT	294.	Cheese

295.	Standardize	308.	0.15-0.60
296.	1.5%	309.	Fat
297.	Action of lipase and proteolytic enzymes	310.	42
298.	Camel	311.	30,000
299.	198.4	312.	375
300.	Phosphatase test	313.	0.15-0.60
301.	Milk solid not fat (MSNF) or lactose	314.	150-200ppm
302.	Turbidity test	315.	135 to150^{0}C without holding
303.	Starter	316.	*Coxiella burnetti*
304.	*Lactobacillus bulgaricus* and *Streptococcus thermophilus.*	317.	4.7
305.	0.68-0.70	318.	1.5
306.	<2 micron	319.	9-11^{0}C
307.	Not less than 28	320.	Acid production

Q. No. 3						
1.T	27.T	53.T	79.T	105.T	131.F	157.T
2.F	28.F	54.F	80.T	106.T	132.F	158.F
3.F	29.T	55.F	81.F	107.T	133.T	159.T
4.T	30.T	56.T	82.T	108.T	134.T	160.F
5.T	31.F	57.T	83.F	109.T	135.T	161.T
6.F	32.T	58.T	84.F	110.T	136.F	162.T
7.T	33.T	59.F	85.T	111.F	137.T	163.T
8.T	34.T	60.T	86.T	112.F	138.T	164.F
9.F	35.F	61.T	87.F	113.T	139.T	165.T
10.T	36.F	62.T	88.F	114.T	140.F	166.T
11.T	37.F	63.T	89.T	115.T	141.F	167.T
12.T	38.T	64.F	90.F	116.F	142.F	168.F
13.F	39.T	65.F	91.F	117.T	143.F	169.F
14.T	40.F	66.F	92.F	118.T	144.T	170.T
15.F	41.F	67.F	93.T	119.T	145.T	171.T
16.T	42.F	68.T	94.T	120.F	146.F	172.F
17.T	43.T	69.F	95.F	121.F	147.F	173.T
18.F	44.F	70.T	96.T	122.F	148.F	174.F
19.F	45.F	71.T	97.T	123.F	149.F	175.T
20.F	46.T	72.T	98.F	124.F	150.F	176.T
21.F	47.F	73.F	99.F	125.F	151.T	177.F
22.T	48.F	74.T	100.T	126.F	152.F	178.T
23.T	49.F	75.F	101.T	127.T	153.T	179.F
24.T	50.F	76.T	102.T	128.T	154.T	180.T
25.F	51.T	77.F	103.F	129.F	155.T	181.T
26.T	52.F	78.F	104.T	130.F	156.F	182.T

183.F	197.F	207.F	221.T	235.T	250.F	264.T
184.T	198.T	208.F	222.T	236.T	251.T	265.T
185.T	199.T	209.F	223.F	237.F	252.F	266.F
186.F	200.F	210.F	224.T	238.F	253.T	267.F
187.T	197.F	211.T	225.T	239.T	254.F	268.T
188.F	198.T	212.T	226.F	240.T	255.T	269.F
189.F	199.T	213.T	227.F	241.T	256.F	270.F
190.F	200.F	214.T	228.T	242.T	257.F	
191.F	201.T	215.F	229.F	243.T	258.T	
192.F	202.F	216.T	230.F	244.T	259.T	
193.F	203.F	217.F	231.F	245.F	260.T	
194.F	204.F	218.T	232.F	246.F	261.T	
195.F	205.T	219.T	233.T	247.F	262.F	
196.F	206.F	220.F	234.F	249.F	263.T	

Q.No. 4			
A		B	
1.	c	1.	h
2.	a	2.	e
3.	b	3.	d
4.	e	4.	g
5.	d	5.	c
6.	f	6.	f
7.	j	7.	b
8.	g	8.	j
9.	h	9.	a
10.	i	10.	i

2

Abattoir and By-products Technology

Q.1 Multiple type questions

1. IS: 4393-1979 of BIS standard deals with
   (a) Code of practice for AM and PM
   (b) Basic requirements for abattoir
   (c) Code of hygienic conditions in meat processing
   (d) None of the above
2. Ideal temperature time combination for scalding broilers is
   (a) 53°C for 120 seconds (b) 60°C for 60 seconds
   (c) 93°C for 5 seconds (d) 60°C for 120 seconds
3. What will be your Post-Mortem decision in case of 'Ascitis' in poultry
   (a) Total unfit for human consumption
   (b) Fit for human consumption
   (c) Partially condemned and passed for human consumption
   (d) None of the above
4. Post mortem judgment for poultry carcass showing lesions of fowl typhoid is
   (a) Total rejection (b) Conditionally passed
   (c) Decision as per carcass state (d) Rejection of local parts
5. Meat unfit for consumption of Jewish people is known as:
   (a) Kosher (b) Schechita
   (c) Terepha (d) None of the above
6. Heparin is obtained from:
   (a) Liver (b) Lungs
   (c) Intestines (d) All of the above

7. One livestock unit is equal to
   (a) One buffalo/cattle (b) Two pigs
   (c) Five sheep/goat (d) All of the above
8. Temperature of scalding of pig is:
   (a) 40°C (b) 50°C
   (c) 60°C (d) 70°C
9. Preslaughter fasting is intended to:
   (a) Improve bleeding (b) Decrease microbial load
   (c) Facilitate dressing (d) All of the above
10. Captive bolt stunning is not recommended in:
   (a) Goat (b) Cattle
   (c) Sheep (d) Pigs
11. V-shaped area in between lairage and slaughter hall is called
   (a) V-hall (b) V-cut slaughter area
   (c) V-race area (d) None of the above
12. In general, the minimum space between the carcasses on the rails should be
   (a) 0.3 to 0.4 m (b) 0.5 to 1 m
   (c) 2 - 5 m (d) >5 m
13. Which of the following is not a meat lymph node of cattle & buffalo
   (a) Prescapular (b) Mediastinal
   (c) Axillary (d) Popliteal
14. HACCP is a
   (a) Comprehensive quality system (b) Comprehensive safety management system
   (c) All of the above (d) None of the above
15. Bones constitute almost ________________% of the weight a dressed carcass
   (a) 20 (b) 15
   (c) 10 (d) 50
16. Casings are mostly obtained from
   (a) Tunica serosa (b) Tunica muscularis
   (c) Tela submucosa (d) Mucous coat

17. Bleeding in cattle should continue for atleast:
    (a) 3 min. (b) 6 min.
    (c) 9 min. (d) 12 min.
18. Cathepsins are __________ in nature.
    (a) Proteolytic (b) Glycolytic
    (c) Lipolytic (d) None of the above
19. Which of the following is not included in pluck of sheep
    (a) Heart (b) Pancreas
    (c) Liver (d) Lungs
20. For effective chilling the speed of air in chilling room should be
    (a) < 0.5 m/s (b) 0.5 to 1 m/s
    (c) 2 - 5 m/s (d) >5 m/s
21. Disease/condition introduced in meat animals during transportation
    (a) Shipping fever (b) Salmonellosis
    (c) Shrinkage (d) All of the above
22. Emergency slaughter is recommended in
    (a) Rabies (b) Anthrax
    (c) Uterine prolapse (d) None of the above
23. A process of insertion of thin steel rod (approx. 0.5 m) in the hole made in skull by the captive bolt pistol is called
    (a) Searching (b) Porging
    (c) Pithing (d) None of the above
24. In post mortem examination of head, masseters are incised to examine for
    (a) *Cysticercus bovis* (b) Echinococcus sps
    (c) Pasteurella sps (d) All of the above
25. Fat surrounding stomach and intestine
    (a) Leaf fat (b) Caul fat
    (c) Pate fat (d) None of the above

26. Organ which can be grouped in edible and also in inedible organs depending on food habits as well as purchasing power of consumers
    (a) Spleen (b) Gall bladder
    (c) Kidney (d) Fetus
27. The blood product that is used as protein reference standard is
    (a) Purified bovine albumin (b) Gamma globulin
    (c) Dried red cell concentrate (d) Crystalline bovine albumen
28. Defect due to keeping the gut in salt for long duration
    (a) Freezer burn (b) Salt burn
    (c) Nodules (d) Nitrate burn
29. ___________ are crushed bones of a size less than 2/232th inch but more than 3/32th inch.
    (a) Bone meal (b) Bone flour
    (c) Bone sinews (d) Bone grists
30. According to the BIS, the specifications for bone meal as livestock feed should contain phosphorus not less than_________% by weight
    (a) 5 (b) 10
    (c) 15 (d) 20
31. Commercially, heparin is isolated either from beef lungs or
    (a) Beef (b) Liver
    (c) Pig intestinal mucosa (d) Kidney
32. Finely ground, dry rendered residue from animal tissue exclusive of hair, hoof, horn, hide trimmings, blood meal, manure and stomach contents is called
    (a) Meat meal (b) Bone meal
    (c) Meat flour (d) Meat and bone meal
33. Poultry by-products meal consists of the ground, dry rendered clean, wholesome parts of the slaughtered poultry exclusive of
    (a) Feathers (b) Gizzard
    (c) Head and legs (d) Feathers and gizzard

34. In preparation of sausages ________coat of the intestines form the actual casings

    (a) Serous (b) Submucus
    (c) Muscular (d) Mucus

35. The protein content of blood meal is in the range of ____________%

    (a) 55-65 (b) 65-75
    (c) 75-85 (d) 85-95

36. Blood meal constitutes about ....of protein

    (a) 85 (b) 70
    (c) 80 (d) 75

37. The processing condition during the dry rendering is

    (a) 40 psi for 4-8 h (b) 80 psi for 4-8 h
    (c) 75 psi for 3-4 h (d) 60 psi for 2 h

38. Casings prepared from cleaned pig stomach is known as

    (a) Rounds (b) Chitterlings
    (c) Middles (d) Maws or stomachs

39. Removal of mucous lining from the casing is known as

    (a) Pulling (b) Sliming
    (c) Fermentation (d) Stripping

40. Meat meal contain about protein and phosphorous

    (a) >55% and <4.5 (b) <55% and <4.5
    (c) <55% and >4.5 (d) >55% and >4.5

41. Verity meat is also known as

    (a) Edible offal's (b) In-edible offal's
    (c) Both of the above (d) None of the above

42. Give your suggestion if someone wants to transport the horses and swine together in a same vehicle

    (a) Shall be transported (b) Shall not be transported
    (c) Divided vehicle can be used (d) None of the above

43. The best way of preserving glands for a long time while retaining its biological activity is
    (a) Salting (b) Sun drying
    (c) Acetone drying (d) None of the above
44. A buffed leather is called
    (a) Full grain (b) Corrected grain
    (c) Natural grain (d) None of the above
45. Sheep and goat casings are generally stored
    (a) Dried (b) Wet and salted
    (c) Raw (d) None of the above
46. Incision should be completed without pause, pressure, stabbing, slanting or tearing in
    (a) Halal method (b) Jewish method
    (c) Jhatka method (d) Evernazine method
47. Pre-slaughter fasting is intended to
    (a) Improve bleeding (b) Decrease microbial load
    (c) Facilitate dressing (d) All of the above
48. As per ICMR recommendations net protein utilization (NPU) of a human being is
    (a) 65 (b) 55
    (c) 60 (d) 70
49. Availability of protein to the man in India is
    (a) 25g/day/head (b) 30g/d/head
    (c) 10g/day/head (d) 40g/d/head
50. Meat for domestic purpose is mainly produced by
    (a) Booth system (b) Line system
    (c) Semi modern abattoirs (d) Modern abattoirs
51. Water pressure requirement for slaughter operations is
    (a) 30 psi (b) 20 psi
    (c) 25 psi (d) 40 psi

52. The storage room temperature where meat is stored till dispatch must be
    - (a) -18 to -22°C
    - (b) 5 to 10°C
    - (c) 0 to -5°C
    - (d) 10-15°C
53. The most preferred rest time for normal animal is
    - (a) 6-12hrs
    - (b) 12-24 hrs
    - (c) 24-36 hrs
    - (d) 36-48hrs
54. Name the online system patented by Canada packers limited Toronto with capacity of 50-150 animals per hours capacity
    - (a) Gravity rail system
    - (b) Intermittent powered system
    - (c) Continuous powered system
    - (d) Canpack system
55. Pluck in pig is not included
    - (a) Spleen
    - (b) Oesophagus
    - (c) Liver
    - (d) Gall bladder
56. Condition which is not included in emergency slaughter is
    - (a) Acute fracture
    - (b) Injuries
    - (c) Uterine prolapse
    - (d) Maggoted wound
57. Edible offal store room must be maintained at
    - (a) 7°C
    - (b) 3°C
    - (c) 10°C
    - (d) 15°C
58. Meat store room must be maintained at
    - (a) 7°C
    - (b) 3°C
    - (c) 10°C
    - (d) 15°C
59. Indicate the time interval in which food and water must be offered during transportation
    - (a) 12 hrs
    - (b) 18 hrs
    - (c) 6 hrs
    - (d) 24 hrs
60. ACTH is released by
    - (a) Anterior pituitary
    - (b) Posterior pituitary
    - (c) Thymus
    - (d) Brain

61. Bruise can be estimated by

    (a) Phytate test (b) Bilirubin test

    (c) Resazurin test (d) None of the above

62. What will be your ante-mortem decision in case of recovered birds from Infectious Laryngotrachitis?

    (a) Unfit for slaughter (b) Passed for slaughter as suspect

    (c) Passed for slaughter as normal (d) None of the above

63. Casualty slaughter is generally performed when

    (a) Animal unnecessary suffering with chronic conditions

    (b) Animal unnecessary suffering with acute conditions

    (c) Both of the above

    (d) None of the above

64. AS per FDA rules ____________ dye is considered safe for carcass identification

    (a) Blue HEL coal tar dye (b) Pink SCH mustard dye

    (c) Purple GL-1 vegetable dye (d) None of the above

65. Trichinoscope is used to identify ____________ in pigs

    (a) *Trichinella spiralis* (b) *Cysticercus bovis*

    (c) *Cysticercus cellulosae* (d) *Echinococcus granulosus*

66. Internal fat in Jewish slaughter is termed as

    (a) Hebrew (b) Cheilev

    (c) Both of the above (d) None of the above

67. Which one is incorrect for time between stunning and sticking?

    (a) <30 seconds for large animals (b) <15 seconds for sheep and goats

    (c) <10 seconds for calves (d) <45 seconds for pigs

68. Elther apparatus is used in ____________ stunning method

    (a) Captive bolt stunning (b) Chemical stunning

    (c) Electrical stunning (d) None of the above

69. Epiphyseal cartilaginous area of cannon bone is called

    (a) Break joint (b) Hit joint

    (c) Knock joint (d) None of the above

70. Jowl cut is found in
    (a) Pigs (b) Buffalo
    (c) Sheep (d) Goats
71. Thymus or pancreas depending on animal age is called
    (a) Bitterbread (b) Sweetbreads
    (c) Poppinbread (d) None of the above
72. Lamb or calf testicles are also called
    (a) Fries (b) Tries
    (c) Dries (d) None of the above
73. Major source of cholesterol in animal body is
    (a) Heart (b) Brain
    (c) Liver (d) Kidney
74. Lecithin is used as
    (a) Emulsifier (b) Antioxidant
    (c) Treatment of snake bite (d) All of the above
75. Sterilized and dehydrated poultry manure contains __________% protein
    (a) 27-29 (b) 30-35
    (c) 10-15 (d) 5-10
76. Nutrient loss is minimum in
    (a) Dry renderer (b) Wet renderer
    (c) Autoclave (d) None of the above
77. Which one is not a tannin bearing plant
    (a) Avarum (*Casia auriculata*) (b) Babul (*Acacia arabia*)
    (c) Myrabalan (*Terminalia chebula*) (d) Neem (*Azadirachta indica*)
78. Saddles are mainly manufactured from
    (a) Vegetable tanned leather (b) Chrome tanned leather
    (c) Aldehyde tanned leather (d) None of the above
79. Glazing is generally applied in
    (a) Vegetable tanned leather (b) Chrome tanned leather
    (c) Aldehyde tanned leather (d) None of the above

80. Technical grade fat is used as
    - (a) Margarine
    - (b) Shortening
    - (c) Cooking fat
    - (d) All of the above
81. Abattoir or meat plant waste is mostly
    - (a) Organic in nature
    - (b) Inorganic in nature
    - (c) Both of the above
    - (d) None of the above
82. Anaerobic system of effluent treatment is most suitable when
    - (a) BOD is >2000 mg/lt
    - (b) High amount of Organic matter
    - (c) Both of the above
    - (d) None of the above
83. Buffed leather is also called
    - (a) Full grain leather
    - (b) Corrected grain leather
    - (c) Both of the above
    - (d) None of the above
84. Quaternary ammonium compounds are good cleaning agent but not preferred in meat industry because
    - (a) Bacteriostatic against G+ve only
    - (b) Inactivated by organic matter
    - (c) Both of the above
    - (d) None of the above
85. Lean or emaciated carcasses have ____________moisture than fat carcasses
    - (a) Higher
    - (b) Lower
    - (c) Almost same
    - (d) None of the above
86. Trickling filters are placed ___________feet deep through which effluent passes
    - (a) 12-15
    - (b) 3-10
    - (c) 1-2
    - (d) None of the above
87. The required water pressure in carcass washing area should be
    - (a) 1000-1700 kPa
    - (b) 500-1000 kPa
    - (c) 2000-25000 kpa
    - (d) None of the above
88. Animal should be provided ____________during rest in lairage
    - (a) Ample amount of water
    - (b) Sufficient feed
    - (c) Proper exercise
    - (d) None of the above

89. Unloading ramp must be with _______degree slop to minimize stress

(a) 20 (b) 30

(c) 40 (d) 10

90. Indicate the conditions arises during transportation of animals

(a) Shipping fever (b) Transit tetany

(c) Salmonellosis (d) All of the above

91. Measly beef condition develops due to

(a) *Cysticercus bovis* (b) *Cysticercus cellulosae*

(c) *Cysticercus cerevisiae* (d) None of the above

92. What will be your decision on meat consumption of still or unborn carcasses in postmortem examination?

(a) Passed for consumption (b) Partial condemnation

(c) Total condemnation (d) None of the above

93. Meat produced by halal procedure is called

(a) Halal meat (b) Zabiha meat

(c) Dhabiha meat (d) All of the above

94. BIS (IS:2537) grading of buffalo carcass is based on

(a) Conformation (b) Finish

(c) Quality (d) All of the above

95. In pork carcass grading, pizzle eye near penis dorsally from aitch bone is found in

(a) Barrow (b) Gilt

(c) Sow (d) None of the above

96. Brain is used to produce

(a) Cholesterol (b) Thromboplastin

(c) Cephalin (d) All of the above

97. Crackling is produced in

(a) Dry rendering (b) Wet rendering

(c) Both of the above (d) None of the above

98. Tankage or slush produced in wet rendering can be utilized as
    - (a) Bedding material
    - (b) Meal
    - (c) Fertilizer
    - (d) None of the above
99. Indicate the true statement related to titre of rendered fat
    - (a) Lard 36-40°C
    - (b) Tallow 42-45°C
    - (c) Sheep fat 44-48°C
    - (d) All of the above
100. Indicate the anticoagulants for blood collection for pharmaceutical purpose
    - (a) Trisodium citrate
    - (b) EDTA
    - (c) Heparin
    - (d) All of the above

Q.2 Fill in the blanks

1. The recommended daily water requirement per pig in abattoir is ______________.
2. The overall intensity of light should not be less than __________________ at all inspection points.
3. ________________ is known as rendered fat of hog.
4. _________________ is a notifiable disease.
5. In anaerobic process of effluent treatment the ponds should be ___________ meter depth.
6. Recovery of fat is better in _________________rendering process.
7. ________________is a white to cream coloured spongy material used in nerve, arterial surgery as well as in prostate operations.
8. __________________________ disease is best judged on ante-mortem examination.
9. ___________________________is a part of hide used for preparation of leather.
10. Dried blood is a rich source of amino acid ________________.
11. The recommended daily water requirement per cattle/buffalo in abattoir is _________litres.
12. The overall intensity of light should not be less than ________________ at all work rooms.
13. PSE condition is normally noticed in ________________.

14. Prolong time between stunning and sticking leads to ________________.
15. Dissolved air floatation method can remove up to ____________% of BOD from meat plant effluent.
16. Walls in the slaughter house should be covered with light color impermeable washable material upto height of ____________________ from floor.
17. In chilling room the minimum space between two adjacent rails for hanging of beef carcass should be ___________________.
18. Neats foot oil is obtained from hooves containing _________________.
19. Most preferred site for construction of an abattoir is_________________.
20. Overscalding of poultry birds leads to ____________________.
21. Ballooning method of hide removal is adopted for __________________.
22. Vertical scalding procedure is considered good in cattle and buffalo due to______________.
23. Best cleaning system for abattoir and meat processing plant is____________.
24. BOD value of abattoir effluent is_________________ppm.
25. The dressing percentage of pigs is _________________.
26. Sweet bread is an another name of _______________________.
27. ____________________ refers to recovery of fat from animal material by heating.
28. Procedure of Jewish slaughter is called____________________.
29. USDA stands for __________________________________.
30. BIS stands for ____.
31. APEDA stands for____ __________.
32. Prayer offers before halal slaughter is called ___________________.
33. Appropriate site for percussive stunning in horned sheep and goat is _____ ____.
34. The period of exposure at 65-70 percent Co_2 concentration in air for pig pre-slaughter anaesthesia is ___________________________.
35. In fabrication of beef cuts right side is called________________________.
36. Dressing percentage is highest in _____________________.
37. PSE stands for______________________.

38. DFD stands for________________________________.
39. Ultimate pH of meat is________________________________.
40. Annual meat production in India is ______________million tones.
41. Meat fit for Jewish slaughter is called _________________.
42. The act of slaughter in Muslim method is called ____________________.
43. Appropriate site for percussive stunning in adult cattle is ____________ ____.
44. Meat cutting room should be maintained at ___________°C with 80% RH.
45. Formula for dressing percentage is____________________________.
46. Among the rail systems, the lowest throughput in____________________ system.
47. Blood meal contains________________% protein.
48. ______________is an enzyme obtained from abomasums of young calf of upto 21 days.
49. ______________is the process of treating the hide with tannins, which displace the water and then combines with the coats and collagen fibers.
50. The yield of animal by-products ranges between____________of live weight.
51. Bones form________________ of the total body weight of the animals.
52. Blood forms____________of the total body weight of the animals.
53. Fish protein concentrate should have at least ___________protein and fat not above 0.75%.
54. Chitin and chitosan are primarily acetylated polymers of_________ with high pH.
55. Glycogen is extracted from_______________obtained as slaughter house waste.
56. Chenodeoxycholic acid found in pig bile is useful in_____________ cholestrol gallstones.
57. The dressing percentage of rabbit meat ranges from____________ in tropical countries.
58. Bleeding of carcass is better in ____________method of slaughter than Jhatka method.

59. Cattle and Sheep bled better in ________________ position , respectively.
60. ________________ is the source of gelatin.
61. Carcasses of males are distinguished from females by____________muscle.
62. Prime grade meat has only a thin layer of external______________.
63. ________________ is must for distribution of meat.
64. ________________ is called sweet bread.
65. ________________ is called weasand.
66. The ultimate pH of meat is depends on________________.
67. A sheep from birth to weaning time is known as________________.
68. Advantage of rabbit as a meat producer as compared to chicken is due to its ________________.
69. Flesh of horse irregularly consumed in ___________ while Whale is consumed in _____________.
70. Sheeps were domesticated as before as________________.
71. Size of domesticated animals is usually ______________than that of their wild ancestors.
72. Wild species of sheep in UK are ________________.
73. Fresians have less proportion of fat in body cavity and ______________ proportion in subcutaneous tissues.
74. Santa Gertrudis consists of ___________ zebu and _____________ shorthorn.
75. ______________was the first breed of pig to be improved scientifically.
76. ________________pig in Hungry has highly pigmented flesh.
77. The most physiologically immature of the three domestic spp. of animals at birth is ____________.
78. The length of gestation period affects weight at ________________.
79. The birth weight of off springs from young mother is ________ than that from mature females.
80. New born pigs are particularly susceptible to sccumb to_____________.
81. Pigs produce _____________ generations in a given period than cattle/ lamb.
82. Electrical stunning uses an alternating current of____________________.

83. The best lard comes from__________________.
84. For bleaching of melted swine fat, _______________________are used as clarifying agents.
85. Fats with a titer of 40°C or higher are called __________________.
86. Hormones oxytocin and vasopressin are extracted from _________________.
87. _________________ enzyme is extracted from rams orchitics or bull testes.
88. ___________________coat of the intestine forms animal casing on processing.
89. Washed and demineralized bone stock should be cooked at_____________ for the manufacture of high quality gelatin.
90. Blood meal contains ________________percent lysine.
91. A byproduct plant is clearly divided into _____________________ sections.
92. Carcass meal contains less than _________________percent protein.
93. Bones contain between ________________ percent ossein.
94. _________________ content are the two factors determining the quality of blood meal.
95. Skin, bones and feet are processed separately from other bones to obtain ______________.
96. The yield of edible by-products from beef, pork and lamb ranges from____________% of the live weight.
97. A___________knife connected to a collection tube is used to obtain the blood in as sanitary a fashion as possible.
98. _______________are used in continuous low temperature rendering system to remove water from liquid mixtures as possible.
99. The best method of preserving most glands to retard autolysis and bacterial growth is by________________.
100. The tough outer membrane of the tongue can be removed more easily by_______________________.
101. Approximately 70% of the bile solids of ruminants is _________________ which are used in the synthesis of cortico- steroids.
102. ____________________________ process is used to obtain maximum efficiency of the acid in ossein production from bones.
103. Extraction of fat from animal tissue by the action of heat is called__________.

104. ________________ is an enzyme extracted from bull testes is called spreading factor as it reduce the viscosity of tissue.

105. The main difference in processing of beef intestine from that of sheep in producing the casings is the intestines are ________________ and then mucus coat is removed.

106. One of the enzyme obtained from blood is________________which is used as a reagent in devices for detection of toxic gasses in industrial operations.

107. Dried blood prepared by special process and reduced to a fine powder is called________________.

108. Cleaned oesophagus is called________________.

109. ________________is one of the most perishable edible by-products.

110. Tripe is obtained from ________________ of cattle.

111. ________________are skins of immature animals.

112. Kidneys are most commonly used ingredients in ________________ food.

113. ________________ are also retail sold as "melt".

114. Lanolin is refined form of ________________.

115. Skin of unborn calf is called________________.

116. Wool is removed from skin by painting of flesh side with________________ and allowing to hang for 24 hrs.

117. According to ISI bone meal must have minmum ________________% calcium.

118. On an average a cow and sheep yield ________________ ft intestines respectively.

119. The blood accounts for ________________% of the body weight of animals.

120. Blood should be collected in ________________ solution immediately after slaughter to obtain blood plasma or albumen.

121. For preservation it is advisable to convert________________ salt of heparin to sodium salt.

122. Pickling of skin is done in salt solution containing ________________ sulfuric acid.

123. All glands must be frozen within________________minutes of killing to be effective for medicinal purposes.

124. Liver is good source of ________________.

125. On an average bones form ________________of the weight of carcass.

126. ______________________ qualities of potable water should be examined.

127. No person suffering from __________________diseases shall be allowed to work in meat factory.

128. An animal not found fit for slaughter shall be marked ______________ and kept in isolation.

129. Carcass infected with anthrax should be _____________ disposed of as per laid down procedures.

130. Formulation of standards provides a ______________ for comparison.

131. Pre-shipment inspection and grading is ____________ for animal casings before export.

132. Animals affected with Tetanus are _______________for slaughter for food.

133. Animals that have recovered from vesicular exanthema with healing lesions are handled as ________________ during slaughter and are not passed for slaughter for food.

134. Animals suffering from demodectic mange which has not progressed to a stage leading to toxemia can be ___________________________________.

135. Animals affected with the condition ochronosis are_______________________________.

136. Strength of sewage or food waste containing organic matter is expressed in terms of __________________.

137. When bacterial pathogen __________________is found in muscle the carcass is passed conditionally for sterilization.

138. Microbiologically animal fats should be regarded with suspicious if they contain more than ________________ percent of free fatty acids.

139. Total bacterial count, presence of _________________and faecal coli indicate the level of hygienic procedures in an abattoir.

140. Leg constitutes around________________ percent of lamb carcass.

141. _________________ is a pale yellow liquid and does not solidified or even at freezing temperature.

142. Blood meal is excellent source of lysine and it constitutes ______________.

143. Average conversion of raw blood to blood meal__________________________.

144. Large abattoir slaughtering >200 livestock units/ day requires___________ acres land area.

145. Recommended intensity of light in inspection points should not be less than____ft candle.

146. Enzyme responsible for tenderization of meat during ageing is __________.

147. Casings are mainly made from________________________layer of intestine.

148. For the preparation of gelatine and glue generally_____________bones are selected.

149. One adult bovine unit is equivalent to ______________sheep or goat.

150. The overall water requirement for pigs is ____________________.

151. Rendered fat of hog is popularly known as __________________.

152. Lairage space should be sufficient for ______________ days supply of cattle and buffalo.

153. Recovery of fat is better in ____________________ rendering process.

154. In general, edible by-products accounts almost ___________ % among all byproducts.

155. Kidney alongwith fat when removed from beef carcass is called __________.

156. Meat can be infested by leaping larvae called____________________.

157. Haemoglobin is a type of ________________protein.

158. Any part of the animal carcass or any material of animal origin not intended for human consumption are called ____________________.

159. Crayons can be obtained by using animal ________________.

160. Yield of the product is better in________________rendering.

Q.3 State whether the following statements are 'True' or False'

1.	Anionic and non-ionic detergents in the ratio of 2:1 are most suitable for meat industry.	( )
2.	The BOD of abattoir effluent is 5 to 6 times higher than domestic sewage.	( )
3.	If more than 10 clear or degenerated cysts are recovered accompanied by emaciation, the carcass is totally condemned.	( )
4.	If Black Quarter is diagnosed at lairage then slaughter is forbidden.	( )
5.	For making the stamp on carcasses, a solution of 1 to 2 % of fuchsine in acetic acid is used.	( )
6.	Specification of Bureau of Indian Standards for evaluation and grading of buffalo carcasses mainly rely on conformation, finish and quality.	( )
7.	Surra is not a notifiable disease in India	( )
8.	Chilling room temperature should not exceed 7^{0}C for carcasses and 3^{0}C for offals.	( )
9.	Lairage space should be sufficient for two days' supply of cattle and three days' supply of sheep or pig.	( )
10.	Ossein is inorganic part of the bone.	( )
11.	Glands should be excised from slaughtered animals within 15-20 minutes after slaughter.	( )
12.	Ovaries are used for insulin production.	( )
13.	Pineal is a source of melatonin hormone.	( )
14.	Water used for cleaning of equipments in meat processing plant should be soft to avoid 'waters spotting'.	( )
15.	Gaseous stunning is recommended in sheep.	( )
16.	If anthrax is diagnosed at lairage then slaughter is forbidden.	( )
17.	Dourine is not a notifiable disease in India.	( )
18.	Beef can be produced in whole of the India.	( )
19.	Poultry birds are generally transported in cooler parts of the day.	( )
20.	Pig gives maximum dressing percentage.	( )
21.	Pithing rod should not be longer than 0.6 meter.	( )
22.	Blood meal contains about 85 percent protein.	( )
23.	Casings are normally obtained from sheep and goat only.	( )
24.	Tanning is not possible in sheep and goat skin.	( )
25.	Chemically glue and gelatine are entirely different.	( )
26.	Bone collagen is called ossein.	( )
27.	Dry rendering produces good quality fat.	( )

28.	Neats foot oil is obtained from scapular bones.	( )
29.	Rennet is generally obtained from suckling goat.	( )
30.	BIS specify the division of Lamb carcass into right and left sides.	( )
31.	Jowl is a cut of buffalo carcass.	( )
32.	In general, Co_2 produces the lowest incidence of PSE.	( )
33.	Pithing is not desirable in present context.	( )
34.	Horizontal bleeding is most desirable posture for buffalo slaughter.	( )
35.	BIS grades for sheep and goats are only six based on conformation, finish and quality	( )
36.	12th rib cross sectional area is having great importance in carcass grading system.	( )
37.	Blood meal is rich in lysine.	( )
38.	Bones obtained from fallen animals are called green bones.	( )
39.	Finishing is referred to fatness of animal which forms important criteria for carcass grading.	( )
40.	Flaying is the skinning operation performed by skilled workers following a uniform pattern.	( )
41.	If the stunning to sticking interval is higher, chances of blood splashing is also increased.	( )
42.	Casings are prepared from muscular layer of intestine of sheep and goat by scraping other layers.	( )
43.	Most efficient bleeding is obtained in halal method among all religious slaughtering methods.	( )
44.	Perfect bleeding refers expulsion of 95% blood from carcass.	( )
45.	Electrical stimulation and stunning can be done simultaneously.	( )
46.	Positioning of electrodes during stunning affects the extent of blood splash in lambs.	( )
47.	Stunning by carbon dioxide method is commonly practiced in pigs.	( )
48.	Stunning improves bleeding of blood in animals after slaughter.	( )
49.	Most of the slaughter houses in India are semi-modern slaughter houses.	( )
50.	Castration of meat animals increases fat deposition in the body.	( )
51.	Meat bone ratio in goats is 3.25:1.	( )
52.	Dressing percentage in sheep is more than goat at the 9 months of age.	( )

53.	The term 'Hog' refers to uncastrated male pig.	( )
54.	Landrace is an important pig breed which has originated in china.	( )
55.	Methyl stilbestrol is a hormone used to manipulate fertility in meat animals.	( )
56.	Semi scald chicken carcasses are preferred for frozen storage.	( )
57.	Practice of domestication of cattle followed establishment of agriculture about 5000BC.	( )
58.	Chinese pigs are derived from species *S. vittatus*.	( )
59.	Trebolone is a steroid used to increase growth of meat animals.	( )
60.	Insulin hormone is commercially extracted from the pancreatic tissue of monkey.	( )
61.	We do not obtain sterilized meal in low temperature rendering.	( )
62.	Heparin is a protein.	( )
63.	Fibrin is removed by constant stirring of coagulated blood.	( )
64.	Blood plasma consists of serum and fibrinogen.	( )
65.	During collection, the glands should be washed with water and then frozen quickly.	( )
66.	Middles are casings prepared from the large intestines of cattle and pigs.	( )
67.	The method of preservation has a definite influence on the weight of hides.	( )
68.	In wet rendering, steam pressure of 20 psi is required for 4 to 6 hrs.	( )
69.	Normal blood clotting time in cattle and pigs is 6 minutes.	( )
70.	Average conversion of raw material to dry meal is 3:1 in dry rendering and 4:1 in wet rendering.	( )
41.	Edible by-products are down graded at time for want of profitable market.	( )
72.	Edible by-products in general are less perishable than carcass meat due to their lesser fat cover.	( )
73.	Sausages containing by-products are often considered nutritionally superior to their all meat counterparts.	( )
74.	Addition of spleen in sausage mixture results in better binding and desirable colour.	( )
75.	Lard has higher titre than tallow.	( )
76.	Smoke point of tallow increases with increase in FFA.	( )
77.	Better quality tallow and higher yield of meal is obtained in wet rendering as compared to dry rendering.	( )
78.	India imports raw hides and skins for its leather industry.	( )

79.	Demineralization is an important step in bone ash production.	( )
80.	In general colour and turbidity of gelatine dilute solution does not influence other properties of usefulness of gelatine.	( )
81	Yield of by-products is relatively greater in sheep and goat compared to cattle and buffaloes.	( )
82.	Tongue and tripe contain relatively higher cholesterol than liver.	( )
83.	Oesophagus is used as sausage ingredient.	( )
84.	Size reduction equipment is essential in the modern continuous rendering system.	( )
85.	Blood is dried in a cooker/drier/conveniently.	( )
86.	Dry rendered meal has lower fat percentage than low temperature rendered meal.	( )
87.	Higher scalding temperature is desired for use of pig skins in leather production.	( )
88.	Sheep skins require a longer time to cool after slaughter as / other hides.	( )
89.	Bacterial degradation is not a problem in bile handling.	( )
90.	Pigs skins deteriorate faster than cattle hides.	( )
91.	Edible and inedible products should be handled in separate rooms.	( )
92.	Post-mortem inspection is limited to examination of viscera only.	( )
93.	Grading of carcasses is generally done by subjective methods.	( )
94.	A reliable and credible meat inspection system exists in Indian slaughter houses.	( )
95.	BOD of effluent from meat processing plants is low.	( )
96.	Hot water should be used for clean up and sterilization purposes in abattoirs.	( )
97.	Carcass cutting and grading practices in India are uniform.	( )
98.	Sheep casings average 25 yards of rounds.	( )
99	Fore saddle and hind saddle of the lamb carcass are separated between 4th and 5th ribs.	( )
100.	Right side of the beef carcass is also known as closed side.	( )
101.	The relative humidity of most cutting room should be maintained around 80 percent.	( )
102	Dry rendering process is advantageous as yield of carcass/ meat meal nearly 20% higher than wet rendering.	( )
103	Bone meal contained about 15% calcium and 32.5% phosphorous.	( )
104.	Best quality of gelatin obtained at 60 ^{O}C processing.	( )

105.	Tallow is obtained from the pig during rendering process.	( )
106.	Glue is chemical similar to the gelatin but inferior in quality.	( )
107.	Bone pieces of less than 2 mm size constitute bone meal.	( )
108.	Lairage should be sufficient to hold 5 days supply of cattle and buffaloes.	( )
109.	Animal must be examined within 48 h of arrival in slaughter house.	( )
110.	Haemorrhagic septicemia is a notifiable disease in India.	( )
111.	Carcase chilling room should be maintained at a temperature of -1 to +3^{0}C.	( )
112.	Porging is a removal of blood vessels from the fore quarter of carcass.	( )
113.	Canpack is a most common line system used in large modern meat plants.	( )
114.	Blood meal is an excellent source of lysine.	( )
115.	If more than 10 clear or degenerated cysts are recovered accompanied by emaciation, the carcass is totally condemned.	( )
116.	If Black Quarter is diagnosed at lairage then slaughter is forbidden.	( )
117.	For making the stamp on carcasses, a solution of 1 to 2 % of fuchsine in acetic acid is used.	( )
118.	Cow shrinks more as compared to bulls and calf during transportation by rail.	( )
119.	Vertical bleeding posture is generally preferred for cattle and buffalo.	( )
120.	Meat, blood and liver meal practically contains no fibre.	( )

Q.4 Match the following

A.

A		B		
1.	Dry rendering	a.	Melt	( )
2.	Green bone	b.	Slaughtered bones	( )
3.	Sweat bread	c.	Sheep intestine	( )
4.	Electrical stimulation	d.	Vertical heating	( )
5.	Wet rendering	e.	Oesophagus	( )
6.	Cat gut	f.	Tenderness	( )
7.	Hyaluronidase	g.	Activated sludge process	( )
8.	Weasand	h.	Thymus	( )
9.	Effluent treatment	i.	Horizontal heating	( )
10.	Spleen	j.	Spreading factor	( )

B.

A		B		
1.	Rib eye cross sectional area	a.	Slaughter procedure	( )
2.	Missed Shock	b.	Pork fat	( )
3.	Schechita	c.	Cattle fat	( )
4.	Lard	d.	Low electrical current	( )
5.	Tallow	e.	Slaughter house	( )
6.	Marbling	f.	12th rib cross sectional area	( )
7.	Searching	g.	Activated sludge process	( )
8.	Back bleeding	h.	Intramuscular fat	( )
9.	Effluent treatment	i.	Adhesions in thoracic cavity	( )
10.	Allana and sons	j.	Lungs contamination with blood	( )

C.

A		B		
1.	Casing	a.	Extended shelf life	( )
2.	Bannuar sheep breed	b.	Submucosa	( )
3.	Jodpi sheep breed	c.	Lean meat	( )
4.	Overhead rail	d.	Carcass quality	( )
5.	Buffalo	e.	Cattle	( )

6.	Leg	f.	Pig	( )
7.	Rib eye	g.	Karnataka	( )
8.	DFD	h.	Captive bold pistol	( )
9.	PSE	i.	Andhra Pradesh	( )
10.	MAP	j.	Modern abattoir	( )

D.

A		B		
1.	Pig feet	a.	Dicalcium phosphate	( )
2.	Demine realization	b.	Steroid drugs	( )
3.	Ossein	c.	Gelling agents	( )
4.	Calcining	d.	Gelatine	( )
5.	Bile	e.	Trotters	( )
6.	Wool grease	f.	Calcitonin	( )
7.	Thyroid	g.	Plasma concentration	( )
8.	Membrane filtration	h.	Haemoglobin bleaching	( )
9.	Hydrogen peroxide	i.	Cholestrol	( )
10.	Pet foods	j.	Bone ash	( )

E.

A		B		
1.	Tripe	a.	Spray drying	( )
2.	Ring drier	b.	Duodenum	( )
3.	Blood plasma	c.	Sausage ingredients	( )
4.	Enterogastrone	d.	Blood	( )
5.	Desert Bone	e.	Neats foot oil	( )
6.	Shin bones	f.	Dead animals bones	( )
7.	Imperfect bleeding	g.	Bilirubin test	( )
8.	Bruising	h.	Emaciation	( )
9.	Bore taint	i.	Malachite green test	( )
10.	Setting and dryout condition	j.	Boiling test	( )

F.

A		B		
1.	Epinaphrine	a.	Calf stomach	( )
2.	Rennet	b.	Pituitary	( )
3.	Tyrypsin	c.	Adrenal	( )

4.	Pork stomach	d.	Variety meat	( )
5.	ACTH	e.	Testes	( )
6.	Calcitonin	f.	Testicles of lamb	( )
7.	Rh factor	g.	Pancrease	( )
8.	Heart	h.	Sausage container	( )
9.	Hyaluronidase	i.	Thyroid	( )
10.	Fries	j.	Bovine albumen	( )

G.

A		B		
1.	Anthrax	a	Negri body test	( )
2.	Brucellosis	b	Colour test	( )
3.	Rabies	c	Identification of adhesions in thoracic cavity	( )
4.	Tuberculosis	d	Ascoli test	( )
5.	Overscalding	e	Bang ring test	( )
6.	Evernnazione	f	Zeil Nielsen staining	( )
7.	Puntilla	g	Jewish slaughtering knife	( )
8.	Rabbinical commission	h	Neck stabbing	( )
9.	Chalef	i	Jewish slaughter	( )
10.	Searching	j	Short double edged knife	( )

H.

A		B		
1.	Caul fat	a	Kidney fat	( )
2.	Cutting fat	b	Pig stomach	( )
3.	Leaf fat	c	Rumen or stomach fat	( )
4.	Oxtail	d	Fat	( )
5.	Red offal	e	Back fat of pork	( )
6.	White offal	f	Variety meat	( )
7.	Grey offal	g	Calf stomach	( )
8.	Dark offal	h	Liver	( )
9.	Rennin	i	Head	( )
10.	Pepsin	j	Intestine	( )

I.

A		B		
1.	Crackling	a	Wound healing	( )
2.	Tankage	b	Bone joints	( )
3.	Feathers	c	Skin of unborn calf	( )
4.	Fibrin powder	d	South American Meat Packers hide	( )
5.	Fibrin foam	e	Dry rendering	( )
6.	Fibrin bioplast	f	East Pakistani hide	( )
7.	Slunk	g	Wet rendering	( )
8.	Frigorifico	h	Leather	( )
9.	Kurbanis	i	Keratin	( )
10.	Corium	j	Arterial and nerve surgery	( )

J.

A		B		
1.	Green bones	a	Fallen animal bones	( )
2.	Desert bones	b	Pig stomach	( )
3.	Isinglass	c	Potteries	( )
4.	Ossein	d	Freshly killed animal bones	( )
5.	Gelatin	e	Shin bones	( )
6.	Glue	f	Caecum	( )
7.	Bone china	g	Dried swim bladder of fish	( )
8.	Neats foot oil	h	Bone protein	( )
9.	Maws	i	Ice-cream	( )
10.	Bung	j	Adhesive	( )

K.

A		B		
1.	Catalase	a	Pineal	( )
2.	Mucin	b	Ovaries	( )
3.	Insulin	c	Testes	( )
4.	Melatonin	d	Gut bread	( )
5.	Thymus	e	Parathyroid	( )
6.	Estrogen	f	Liver	( )
7.	Hyaluronidase	g	Stomach	( )
8.	Bile	h	Pancreas	( )
9.	Thyrocalcitonin	i	Gall bladder	( )
10.	Pancreas	j	Thyroid	( )

Key

Abattoir & Byproducts Technology

Q.No. 1			
1.b	31.c	61.b	91.a
2.a	32.d	62.b	92.c
3.a	33.d	63.a	93.d
4.a	34.b	64.c	94.d
5.c	35.c	65.a	95.a
6.d	36.c	66.c	96.d
7.d	37.c	67.d	97.a
8.c	38.d	68.b	98.b
9.d	39.b	69.a	99.d
10.b	40.c	70.a	100.d
11.c	41.a	71.b	
12.a	42.b	72.a	
13.b	43.c	73.b	
14.b	44.b	74.d	
15.a	45.b	75.a	
16.c	46.b	76.a	
17.b	47.d	77.d	
18.a	48.a	78.a	
19.b	49.c	79.a	
20.c	50.a	80.d	
21.d	51.b	81.a	
22.c	52.a	82.c	
23.c	53.b	83.b	
24.a	54.d	84.c	
25.b	55.d	85.a	
26.a	56.d	86.b	
27.d	57.b	87.a	
28.b	58.a	88.a	
29.d	59.a	89.a	
30.c	60.a	90.d	

Q.No. 2	
1.	454lt/d/pig
2.	540 lux or 50 ft candle
3.	Lard
4.	Anthrax
5.	3-5
6.	Wet
7.	Fibrin foam
8.	Tetanus/rabies/poisonings
9.	Dermis or corium
10.	Lysin
11.	272
12.	220lux/20 ft candle
13.	Pigs
14.	Muscle haemorrhages or blood splashing
15.	75-94%
16.	Atleast 3 meters
17.	0.9m^2
18.	Shin Bones
19.	Semi Urban
20.	Discolouration
21.	Sheep and goats
22.	High bleeding
23.	Central cleaning system
24.	1400-3200
25.	80%
26.	Thymus
27.	Rendering
28.	Schechita
29.	United States Department of Agriculture
30.	Bureau of Indian Standards
31.	Agricultural & Processed Food Products Export Development Authority
32.	Bismi

33.	Top of head behind the ridge connecting two horns and directed towards gullet
34.	45 seconds
35.	kidney close side
36.	Pigs
37.	Pale Soft and Exudative
38.	Dark Firm and Dry
39.	5.6
40.	7.3
41.	Kosher
42.	Halal
43.	Middle of forehead where two lines drawn from medial canthus of each eye to the base of opposite horn or horn prominence cross
44.	15
45.	Carcass bwt/Live wt x100
46.	Gravity rail
47.	>80
48.	Rennin
49.	Tanning
50.	50-60%
51.	12-28%
52.	5-7%
53.	67.5%
54.	Glucosamine
55.	Alpha cells of pancrease
56.	Dissolving
57.	50-52%
58.	Halal
59.	Vertical and horizontal
60.	Bone
61.	Pizzle
62.	Fat
63.	Cold chain
64.	Thymus or pancrease
65.	Oesophagus

66.	Glycogen
67.	Lamb
68.	High muscle bone ratio
69.	Denmark
70.	7000 BC
71.	Smaller
72.	Soay and Shetland
73.	Less
74.	3/8 & 5/8
75.	Landrace
76.	Mangalista
77.	Pigs
78.	Parturition
79.	Less
80.	Stress
81.	More
82.	0.3 amperes and 70-90volts for 2-10 seconds
83.	Caul and Kidneys
84.	Fuller's earth or activated carbon
85.	Tallow
86.	Posterior pituitary gland
87.	Hyaluronidase
88.	Submucosa
89.	60^0C
90.	6-9
91.	Clean and unclean
92.	55
93.	33- 36
94.	Moisture and ammonia
95.	Neats foot oil
96.	20-30
97.	Hollow
98.	Evaporation
99.	Quick freezing
100.	Blanching/dipping in boiling water

101.	Cholic and deoxycholic acid
102.	Counter current demineralization
103.	Rendering
104.	Hyalurinidas
105.	Turned inside out
106.	Cholinesterase
107.	Blood floor
108.	Weasand
109.	Liver/offal
110.	Rumen and reticulum
111.	Kip
112.	Pet
113.	Spleen
114.	Wool grease
115.	Slunk
116.	Sodium sulfide
117.	32
118.	140 ft and 80 ft
119.	8%
120.	4% trisodium citrate
121	Barium
123.	30
124.	Iron and vitamin-A
125.	15%
126.	Chemical and bacteriological
127.	Contageous
128.	Suspect
129.	Immediately
130.	Basis
131.	Compulsory
132.	Unfit
133.	Suspect
134.	Passed after removal of diseased parts
135.	Passed after removing the affected parts, if not generalized
136.	BOD

137.	Erycephalothrix
138.	Two
139.	Salmonella
140.	32
141.	Neat Foot Oil
142.	6-9%
143.	1/5th of blood weight (5:1)
144.	4-6
145.	50
146.	Cathepsin
147.	Submucous coat/Telasubmucosa
148.	Long bones
149.	5
150.	472 lt/pig with 10% extra
151.	Lard
152.	3
153.	Wet
154.	67%
155.	Kidney Knob
156.	Skipper
157.	Metaloprotein
158.	Animal by-products
159.	Fat
160.	Dry

Q.No. 3		
1. T	41.T	81.T
2.T	42.F	82.F
3.T	43.T	83.T
4.T	44.F	84.T
5.T	45.F	85.F
6. T	46.T	86.F
7.F	47.T	87.F
8.T	48.F	88.T
9.F	49.F	89.T
10.T	50.T	90.F
11.T	51.T	91.T
12.F	52.T	92.F
13.T	53.F	93.T
14.T	54.F	94.F
15.F	55.F	95.F
16.T	56.T	96.T
17.T	57.T	97.F
18.F	58.T	98.F
19.T	59.T	99.F
20.T	60.F	100.T
21.T	61.T	101.T
22.T	62.F	102.T
23.F	63.F	103.F
24.F	64.T	104.T
25.F	65.F	105.F
26.T	66.T	106.T
27.F	67.T	107.T
28.F	68.F	108.F
29.F	69.T	109.F
30.F	70.T	110.T
31.F	71.T	111.T
32.T	72.F	112.T
33.T	73.T	113.T
34.F	74.F	114.T
35.T	75.F	115.T
36.T	76.F	116.T
37.T	77.F	117.T
38.T	78.T	118.T
39.T	79.F	119.T
40.T	80.F	120.T

Q.No. 4											
A		B		C		D		E		F	
1.	i	1.	f	1.	b	1	e	1.	c	1.	c
2.	b	2.	d	2.	g	2.	a	2.	d	2.	a
3.	h	3.	a	3.	i	3.	d	3.	a	3.	g
4.	f	4.	b	4.	j	4.	j	4.	b	4.	h
5.	d	5.	c	5.	h	5.	b	5.	f	5.	b
6.	c	6.	h	6.	c	6.	i	6.	e	6.	i
7.	j	7.	i	7.	d	7.	f	7.	i	7.	j
8.	e	8.	j	8.	e	8.	g	8.	g	8.	d
9.	g	9.	g	9.	f	9.	h	9.	j	9.	e
10.	a	10.	e	10.	a	10.	c	10.	h	10.	f

G		H		I		J		K	
1.	d	1.	c	1.	e	1	d	1.	f
2.	e	2.	e	2.	g	2.	a	2.	g
3.	a	3.	a	3.	i	3.	g	3.	h
4.	f	4.	f	4.	a	4.	h	4.	a
5.	b	5.	h	5.	j	5.	i	5.	j
6.	h	6.	d	6.	b	6.	j	6.	b
7.	j	7.	J	7.	c	7.	c	7.	c
8.	i	8.	i	8.	d	8.	e	8.	i
9.	g	9.	g	9.	f	9.	b	9.	e
10.	c	10.	b	10.	h	10.	f	10.	d

3

Fresh Meat Technology

Q.1 Multiple type questions

1. India contributes______________percent in world meat production.

    a. 2   b. >2

    c. 1   d. <1

2. Functional unit of skeletal muscle is

    a. Myofibril   b. H-zone

    c. M-line   d. Sarcomere

3. Water held by capillary forces in meat is called

    a. Free water   b. Immobilized water

    c. Bound water   d. None of the above

4. Name the meat required high oxygen in modified atmospheric packaging to maintain bright red colour

    a. Pork   b. Poultry meat

    c. Chevon   d. Beef

5. The permissible limit of nitrate and nitrite as preservative in meat products is

    a. 500 and 200 ppm   b. 200 and 200 ppm

    c. 200 and 100 ppm   d. None of the above

6. Green discolouration in meat is due to

    a. *Clostridium putrefaciens*   b. *E.coli*

    c. *Pseudomonas mephitica*   d. None of the above

7. Product orientation evaluation of meat products in sensory evaluation process is done by

    a. Consumer panel   b. Specialized panel

    c. Both of the above   d. None of the above

8. Name extensively used serological method for meat species speciation
   a. Double immuno diffusion test
   b. Precipitation or ring test
   c. Single radial immunodiffusion test
   d. None of the above
9. "Rapka" is a popular traditional meat product belongs to:
   a. Arunachal Pradesh
   b. Jammu & Kashmir
   c. Himachal Pradesh
   d. Madhya Pradesh
10. Which one is not a connective tissue protein:
    a. Collagen
    b. Reticulin
    c. Elastin
    d. Myosin
11. Name the carbohydrate usually found in muscle at the time of slaughter:
    a. Glycogen
    b. Glucose
    c. Galactose
    d. Sucrose
12. Meat colour is a total visual perception of meat and expressed in terms of:
    a. Hue
    b. Chroma
    c. Value
    d. All of the above
13. Meat is considered as:
    a. Low acid food
    b. High acid food
    c. Medium acid food
    d. None of the above
14. Which is not applicable in smoking preservation:
    a. Surface dehydration
    b. Freezer burn
    c. Antioxidant property
    d. Lowering of surface pH
15. ISO-9000 is a:
    a. Food safety system
    b. Detailed quality management system
    c. Food Sanitary system
    d. None of the above
16. Bacon is obtained from
    a. Pork
    b. Beef
    c. Chevon
    d. Mutton

17. Right side of carcass in beef is called:
    a. Kidney close
    b. Kidney free
    c. Kidney knob
    d. None of the above
18. MFPO was enforced in the year
    a. 1973
    b. 1983
    c. 1993
    d. 2000
19. The commonly used frequencies in microwave cookery are
    a. 950 &3450 MHz
    b. 915 &2040 MHz
    c. 815 & 2450 MHz
    d. 915 & 2450 MHz
20. RDA of protein in adult human being is
    a. 56g
    b. 100g
    c. 32g
    d. None of the above
21. The intra-muscular fat is highest in
    a. Beef
    b. Poultry meat
    c. Mutton
    d. Chevon
22. Which is not a method of moist heat cookery?
    a. Stewing
    b. Braising
    c. Swissing
    d. Broiling
23. Acceptable weight loss during smoking is
    a. 5-10 %
    b. 10-15%
    c. 15-20%
    d. 20-30%
24. Onset of rigor mortis in pork occurs between
    a. 6-12 hrs of postmortem
    b. 5 minutes to 1 hour of postmortem
    c. 15 minutes to 3 hours of postmortem
    d. None of the above
25. Flesh becomes white on cooking in
    a. Veal
    b. Pork
    c. Beef
    d. None of the above

26. The a_w of intermediate food is
    a. 1-1.5
    b. 0.6-0.85
    c. 0.95-1.0
    d. None of the above
27. The bound water in meat is
    a. 4-6%
    b. 10-12%
    c. 15-20%
    d. None of the above
28. Titin is
    a. Regulatory protein
    b. Contractile protein
    c. Inhibitory protein
    d. Cytoskeletal protein
29. The most abundant fatty acid in animal body is
    a. Oleic acid
    b. Linoleic acid
    c. Linolenic acid
    d. Palmitic acid
30. Meat is poor source of
    a. Iron
    b. Calcium
    c. Sulphur
    d. Potassium
31. Stable pink colour of cured meat products is due to
    a. Myoglobin
    b. Metmyoglobin
    c. Nitrosomyoglobin
    d. Nitrosohaemochromogens
32. The most abundant amino acid present in collagen is
    a. Desmosine
    b. Glycine
    c. Proline
    d. Lysine
33. One sarcomere length in a myofibril is the distance between
    a. Two adjacent Z-line
    b. Two adjacent A-bands
    c. A adjacent M-line
    d. Two adjacent H-zones
34. Irradiation doses used in Radappertization of meat ranges from
    a. 30-40 KGy
    b. 80-1000KGy
    c. 1-5KGy
    d. 50-60 KGy
35. Which one is a contractile protein
    a. Myosin
    b. Tropocollagen
    c. Troponin
    d. M-protein

36. Status of glycogen (%) at the time of slaughter in muscles is

    a. 0.3 b. 0.6

    c. 1.0 d. 0.8

37. ______________freezing favours formation of pure ice crystals and concentration of solutes in unfrozen solution.

    a. Slow b. Fast

    c. All of the above d. None of the above

38. Colour of fresh beef is

    a. Greyish pink b. Brownish pink

    c. Gray-white d. Bright cherry red

39. Chemical reactions causing oxidative rancidity is called

    a. Dehydration b. Shrinkage

    c. Autooxidation d. Discolouration

40. Illness caused by ingestion of toxins is called

    a. Food infection b. Food poisoning

    c. Contamination d. None of the above

41. Meat is an especially good source of

    a. Calcium b. Iron

    c. Copper d. Selenium

42. Cathepsin is a proteolytic enzyme used in tenderization of meat is found in

    a. Golgi- complex b. Mitochondria

    c. Lysosomes d. T-tubules

43. Meat is a good source of Vitamin

    a. B-Complex b. Vitamin –C

    c. Folacin d. None of the above

44. Protein contributes about____________% of total muscle mass

    a. 16-22 b. 25-30

    c. 10-15 d. 5-10

45. Which term is related to meat production
    a. Grey revolution
    b. White revolution
    c. Pink revolution
    d. Yellow revolution
46. Indian buffalo meat is having strong demand in International market due to
    a. High lean and fat
    b. High lean and less fat
    c. Lean and organic meat
    d. Lean and inorganic meat
47. Which Municipal slaughter house has been approved for exporting chevon to UAE
    a. Mumbai
    b. Hyderabad
    c. Chennai
    d. Delhi
48. Calpain slowly disrupt the Z-line in tenderization process by release of
    a. β-actinin
    b. α-actinin
    c. γ- actinin
    d. Desmosin
49. The articular cartilage of adjacent bone joint together contain a specialized connective tissue substance called
    a. Ground substance
    b. Seminiferous tubules substance
    c. Mitochondrial substance
    d. Synovial fluid
50. Name the muscle in which pH decline is quickest during conversion of muscle to meat than other muscles of the animal carcass
    a. Semimembronosus
    b. Semispinalis
    c. Psoas
    d. Seratus ventralis
51. India is having around --------- state of art mechanised abattoir.
    a. 50
    b. 30
    c. 100
    d. Zero
52. Collagen can be estimated in terms of amino acids
    a. Lysine
    b. Tyrosin
    c. Hydroxyproline
    d. None of the above
53. The porphyrin ring in myoglobin contains metallic ion of
    a. Magnesium
    b. Copper
    c. Iron
    d. None of the above

54. The ATP splitting activity of myosin ATP are greatly enhanced by the
    a. Calcium ion released into sarcoplasm
    b. Sodium ion released into sarcoplasm
    c. Potessium released into sarcoplasm
    d. None of the above
55. The most common food pathogens are inhibited at an a_w of
    a. 0.85
    b. 0.88
    c. 0.91
    d. 0.92
56. Low acid food products have a pH of
    a. >3.9
    b. >3.6
    c. >4.6
    d. >4.9
57. Eugenol, an antibacterial substance is present in
    a. Onion
    b. Cinnamon
    c. Clove
    d. Bay leaf
58. The minimum radiation doses on Mrad for the radappertization of pork is
    a. 2.3
    b. 4.7
    c. 3.7
    d. 5.1
59. The most important psychrotrophic bacteria in meats belong to the genus
    a. Pseudomonas
    b. Salmonella
    c. Flavobacterium
    d. Mucor
60. The lowest recorded temperature of growth for bacterium is
    a. -10°C
    b. -20°C
    c. -30°C
    d. -45°C
61. It is harder to kill those bacterial cells through heat treatment have
    a. High lipid content
    b. High amino acid content
    c. High water content
    d. High mineral content
62. During low temperature storage (chilling) enzymatic and microbial changes in food are
    a. Hastened
    b. Prevented
    c. Slowed down
    d. None of the above

63. Thermal death point is the temperature necessary to kill all the organism in
   a. 20 minutes
   b. 15 minutes
   c. 25 minutes
   d. 10 minutes
64. *Bacillus coagulans* is a
   a. Thermoduric organism
   b. Thermophilic organism
   c. Psychrophilic organism
   d. Osmophilic organism
65. The polypeptides in collagen are
   a. Three stranded helical structures
   b. Two stranded helical structures
   c. Four stranded helical structures
   d. None of the above
66. Many myosin molecules assemble together to form the
   a. Thin filaments of muscle
   b. Thick filaments of muscle
   c. Endomysium
   d. None of the above
67. When collagen is heated in water to 80°C
   a. Collagen remains insoluble
   b. Collagen begins to be converted into gelatin
   c. Collagen fibres gets only shortened
   d. None of the above
68. Red muscle fibres
   a. Lacks mitochondria
   b. Rich in mitochondria
   c. Rich in haemoglobin
   d. None of the above
69. Desmin is a
   a. Z-line protein
   b. M-line protein
   c. C-line protein
   d. None of the above
70. Structure which are seen in the space and develop in muscles when stretched beyond the point of overlap of A&I filaments are called
   a. Gap filaments
   b. CASF
   c. Thick filaments
   d. None of the above
71. Nebulin is a
   a. M-line protein
   b. N-line protein
   c. H-line protein
   d. None of the above

72. Hydroperoxides formes during auto-oxidation of fats are
    a. Very stable and do not break up further
    b. Very unstable and break up further giving rise to aldehyde, acids or ketones
    c. Both of the above
    d. None of the above
73. Formation of gel from gelatin is
    a. Reversible on heating
    b. Irreversible on heating
    c. It again changes into collagen like structure
    d. None of the above
74. The membrane potential in muscle is the result of
    a. The active transport of ions through the membrane
    b. Selective permeability of the membrane to the diffusion of ions and molecules
    c. Unique ionic composition of the intracellular and extra cellular fluids
    d All of the above
75. Fresh meat requires a packaging material which is
    a. Impermeable to oxygen
    b. Permeable to moisture
    c. Transparent to light
    d. Heat shrinkable
76. Bright red colour of fresh meat is due to
    a. Myoglobin
    b. Oxymyoglobin
    c. Nitrosomyoglobin
    d. Metmyoglobin
77. Bacterial spoilage of fresh meat is accelerated in the pH range of
    a. 3.5-4.5
    b. 4.5-5.0
    c. 5.3-6.0
    d. None of the above
78. Pigment responsible for brown colour of meat is
    a. Myoglobin
    b. Metmyoglobin
    c. Nitrosomyoglobin
    d. Oxymyoglobin
79. Optimum relative humidity in refrigeration preservation of meat is
    a. 88-92
    b. 94-95
    c. 80-85
    d. None of the above

80. Role of nitrate and nitrite in meat preservation as a curing ingredient
    a. Colour development
    b. Flavour enhancer
    c. Bactericidal
    d. All of the above
81. The fresh meat and fish has a water activity
    a. >0.95
    b. 0.91
    c. 0.88
    d. <0.80
82. Major share of meat in export is
    a. Deboned &deglanded frozen Carabeef
    b. Chilled & frozen chevon
    c. Chilled & frozen poultry meat
    d. Chilled & frozen pork
83. Which one is not a myofibrillar protein
    a. Myosin
    b. Actin
    c. C-protein
    d. Myoglobin
84. Hurdle concept was coined by
    a. Leistner and Rodel
    b. Flemmings
    c. Nicolus appert
    d. None of the above
85. Green rot in egg is caused by
    a. *Serratia marcescens*
    b. Enterobactor
    c. *Pseudomonas fluorescence*
    d. *Flavobacterium ctyophaga*
86. Eugenol, an antimicrobial substance is present in
    a. Onion
    b. Cinnamon
    c. Garlic
    d. Clove
87. Enzyme used for tenderization of meat is
    a. Papain
    b. Ficin
    c. Bromelin
    d. All of the above
88. Which among the following is not a meat based bioactive compound
    a. Taurine
    b. Carnitine
    c. CLA
    d. Histidine

89. ISO-22000 is a

    a. Food quality management system

    b. Food safety management system

    c. Environmental management system

    d. All of the above

90 …………is used to retain curing solution in product during curing.

    a. Ascorbate
    b. Nitrate
    c. Phosphate
    d. BHT

91 ……………………is used for the packaging of large and uneven cuts.

    a. Vacuum packaging
    b. Modified atmospheric packaging
    c. Shrink packaging
    d. Intelligent packaging

92. ICMR recommendation for meat availability in India is

    a. 30g/head/day
    b. 50g/head/day
    c. 80g/head/day
    d. 100g/head/day

93. In India, only______% of total meat produced is converted into value added products

    a. 5
    b. 10
    c. 2
    d. 12

94. Highest share of meat in Indian export market is

    a. Poultry meat
    b. Mutton
    c. Buffalo meat
    d. Pork

95. FSSAI was implemented in the year

    a. 2011
    b. 2018
    c. 2005
    d. 2000

96. Why India is exported deglutinated or deboned meat

    a. FMD virus localization in spines
    b. High demand
    c. To reduce the weight
    d. All of the above

97. Connective tissue sheath surrounding the entire muscle is

    a. Perimysium
    b. Epimysium
    c. Endomysium
    d. None of the above

98. Muscle fibres makes about____________% volume of muscle
    a. 75-92
    b. 60-75
    c. 50-60
    d. None of the above
99. Myosin protein is responsible for
    a. Water holding capacity
    b. Emulsifying capacity
    c. Meat tenderness
    d. All of the above
100. Which one is true for completion phase of rigor mortis?
    a. Muscle becomes relatively inextensible
    b. Shorten sarcomere length
    c. Both of the above
    d. None of the above
101. What is effect of cold shortening on I-band
    a. Disappears completely
    b. No complete disappearance
    c. No effect on I-band
    d. All of the above
102. What is effect of thaw rigor on I-band
    a. Disappears completely
    b. No complete disappearance
    c. No effect on I-band
    d. All of the above
103. The most abundant neutral fat in meat is
    a. Triglycerides
    b. Diglycerides
    c. Monoglycerides
    d. None of the above
104. Meat is good source of bioactive peptides like
    a. Taurine
    b. Carnitine
    c. CLA
    d. All of the above
105. Approximate quantity of carotene found in beef is
    a. 0.14-0.225 mg/g fat
    b. 0.25-0.552mg/g fat
    c. 0.50-0.615mg/g fat
    d. None of the above
106. Which character is not true for fast freezing?
    a. Less drip loss
    b. Less muscular damage
    c. Less shrinkage
    d. Large ice crystal formation

107. Indicator organism for canning is

a. *Clostridium botulinum*
b. *Coxiella burnetii*
c. *Streptococcus lactis*
d. None of the above

108. Irradiation sterilization dose to achieve radappertization is

a. 100-120kGy
b. 30-40 kGy
c. 50-80kGy
d. 80-100kGy

109. Irradiation pasteurization dose to achieve radicidation is

a. 2.5-10 kGy
b. 1-2 kGy
c. 10-15 kGy
d. 15-20 kGy

110. Irradiated food must be labelled with International symbol

a. Santura
b. Pasturema
c. Redura
d. None of the above

111. Commonly used antibiotics in meat preservation is

a. Nisin
b. Natamycin
c. Both of the above
d. None of the above

112. Commonly used antibiotics in fish preservation is

a. Aureomycin
b. Terramycin
c. Chloromycetin
d. All of the above

113. Commonly used enzyme for tenderization of meat is

a. Papain
b. Ficin
c. Bromelin
d. All of the above

114. Which is true for electrical stimulation?

a. Advances the rigor mortis
b. pH 6.0 achieved within 2-3 hrs
c. Both of the above
d. None of the above

115. Artery pumping for curing is limited to

a. Ham
b. Brisket
c. Round
d. None of the above

116. Compound of smoke responsible for skin formation in sausages is

a. Alcohol
b. Carbonyls
c. Organic acid
d. None of the above

117. Which among is not the method of moist heat processing

a. Stewing
b. Braising
c. Simmering
d. Roasting

118. Which is the example of fat mimetic?

a. Simplesse
b. Dairylo
c. Both of the above
d. None of the above

119. Fat substitutes used in meat products are

a. Olestra
b. Saltrim
c. Caprenin
d. All of the above

120. Compound can be utilized as oxygen scavengers in meat products is

a. Glucose oxidase
b. Rice extract
c. Immobilized yeast
d. All of the above

121. Name the example of humectant

a. Glycerine
b. Sugar
c. Some salts
d. All of the above

122. Elastin contains two unique amino acid________________ present @eight amino acids/1000 amino acids residues.

a. Desmosine
b. Isodesmosine
c. Both of the above
d. None of the above

123. Myosin ATPase is activated by the Ca++ ions, whereas, Actomyosin ATPase is activated by ions

a. Mg++
b. Ca++
c. Both of the above
d. None of the above

124. Name the species is associated with the spoilage of canned meats

a. Bacillus
b. Clostridium
c. Both of the above
d. None of the above

125. Indicate the true statement for nutritionally speaking, organ meats

a. Low in fat
b. High in Vitamins
c. High in minerals
d. All of the above

126. Name the major emulsifying proteins present in sausages emulsion

a. Actin
b. Myosin
c. Both of the above
d. None of the above

127. Indicate the sweeteners commonly used in sausage making

a. Sucrose
b. Dextrose
c. Corn syrup
d. All of the above

128. Saran is widely used packaging of meat products is made up of polymers like

a. Vinylidine chloride
b. Vinyl chloride
c. Both of the above
d. None of the above

129. Indicate the right combination of enzyme and its source used for tenderization of meat

a. Papain- Papaya
b. Bromelin-Pineapple
c. Ficin- Figs
d. All of the above

130. In sausage emulsion, the term binder encompasses

a. Water binding
b. Fat emulsification
c. Both of the above
d. None of the above

131. For the preparation of fermented sausage, starter culture of ________ is commonly used

a. Lactobacillus
b. *Pediococcus cerevisae*
c. Both of the above
d. None of the above

132. Name the freeze dried products which have attained commercial success.

a. Chops
b. Steaks
c. Both of the above
d. None of the above

133. Commonly accepted important theories of olfaction are

a. Societal-reaction theory
b. Vibrational theory
c. Enzymatic theory
d. All of the above

134. ____________ is used to measure the concentration of salt in solution

a. Salinometer
b. Conductivity meter
c. Thermometer
d. None of the above

135. ____________ is used to measure the total soluble solids in solution

a. Salinometer
b. Conductivity meter
c. Thermometer
d. None of the above

136. The write statement for loss in canning is

a. Vitamin-B_2 and Niacin 10%
b. Biotin 20%
c. Pantothenic acid 25%
d. All of the above

137. Indicate the true statement for surimi preparation

a. Removal of water soluble proteins
b. Addition of cryoprotectant
c. Both of the above
d. None of the above

138. Troponin is globular protein complex containing ________ sub units

a. T
b. I
c. C
d. All of the above

139. Hydration of meat in maximum at ionic strength of 0.8 to 1.00 while phosphates reduce the salt concentration required to produce maximum

a. Swelling
b. Drip loss
c. Browning
d. None of the above

140. Indicate the high and low range of pulsed voltages applied for muscles electrical stimulation

a. High 500-1000V & Low Ca 90V
b. High 100-1500V & Low Ca 70V
c. High 100-500V & Low Ca 80V
d. None of the above

141. Indicate the distinct disadvantage of aluminum foil as package

a. High cost
b. Low strength
c. Both of the above
d. All of the above

142. Reddish colour fluid of fresh meat

a. Weep
b. Drip
c. Shrink
d. None of the above

143. Indicate the true proportion of meat portion of animal body

a. 70% in myofibrils
b. 20% in the sarcoplasma
c. 10% in the connective tissue
d. All of the above

144. Manipulation of muscle mass can be achieved by

a. Nutrition
b. Gutpeptides
c. Transgenensis
d. All of the above

145. Marbling is seen in

a. Beef
b. Pork
c. Both of the above
d. All of the above

146. Starting with the innermost, connective tissues involved in muscle structure are

a. Sarcolemma, endomysium, epimysium and perimysium
b. Sarcolemma, perimysium, endomysium, and epimysium
c. Both of the above
d. None of the above

147. Cathepsins are responsible for changes in

a. Cell membrane permeability
b. Protein structure.
c. Both of the above
d. All of the above

148. Name the amino acids required for including a protein in collagen class

a. Hydroxy proline
b. Hydroxyl lysine
c. Both of the above
d. All of the above

149. Collagen is built up of triplets of amino acids containing

a. Glycine, proline and hydroxyl proline
b. Glycine, toluene and hydroxyl toluene
c. Proline, hydroxyl proline and titin
d. None of the above

150. The inner mitochondrial membrane contains enzymes of

a. Electron transport chain
b. ATPase
c. Both of the above
d. All of the above

151. Indicate the protein component of nucleoproteins

a. Protamines
b. Histamines
c. Both of the above
d. All of the above

152. Most gel in water show vapour pressure and electric conductivity very close to

a. Water
b. Ice
c. Solid
d. None of the above

153. Correct diameter and length of thick filaments of vertebrates is

a. 14-16nm &1.5µm respectively
b. 10-12nm &1.0µm respectively
c. 17-20nm &2.5µm respectively
d. None of the above

154. Fungi of genera __________is the only bacterium of public health importance that can grow at a_w values near 0.86.

a. Acetobacter
b. Xerophillic Fungi
c. Both of the above
d. None of the above

155. Sodium benzoate acts essentially as______________inhibitor

a. Mold
b. Yeast
c. Both of the above
d. None of the above

156. Indicate the polypeptide antibiotic

a. Nicin
b. Subtilin
c. Both of the above
d. None of the above

157. The skeletal muscle myosin contains unusual amino acids such as

a. 3-methyl histidine
b. N-trimethyl lysine
c. N- monomethyl lysine
d. All of the above

158. Indicate the correct combination of important flavour producing compound from carbohydrates due to low of water molecules on heating

a. Furfurol from pentoses
b. Hydroxy methyl furferol from hexoses
c. Both of the above
d. None of the above

159. The polypeptide chain of the primary structure of collagen has repeating sequence of

a. Glycine, proline and hydroxyproline
b. Alenine, leusine and isoleucine
c. Glycine, methionine and tryptophane
d. None of the above

160. An emulsion is two phase system consisting of

a. Continuous and dispersed phase b. Solid and liquid phase

c. Discontinuous and dilute phase d. All of the above

161. On tryptic digestion myosin filament splits into

a. H. meuromyosin b. L-meuromyosin

c. Both of the above d. None of the above

162. Time in minutes to kill 90% micro-organisms of food at a given temperature is called

a. D-value b. Z-value

c. T-value d. None of the above

163. Increase in degree temperature to obtain 10 fold increases in death rate of microbes is called

a. D-value b. Z-value

c. T-value d. None of the above

164. Time required in minutes at 121°C to destroy spores or vegetative cells of specific organism is called

a. D-value b. Z-value

c. F-value d. None of the above

165. Creation of ____________ is an important step in production of sectioned and formed meat

a. Protein matrix b. Fat emulsion

c. Protein-fat matrix d. None of the above

166. Meat emulsion is categorized as____________ emulsion

a. Oil in water b. Water in oil

c. Both of the above d. None of the above

167. Biobased food packaging materials are derived from

a. Renewable sources b. Anaerobic microbes

c. Both of the above d. None of the above

168. Moat suitable packaging material for large and uneven cuts of fresh meat is

a. MAP b. CAP

c. Shrink film overwarap d. None of the above

169. Hedonic scale is of Greek origin and relates to

a. Degree of like and dislike

b. Degree of acceptance and rejection

c. Degree of freedom to select the product

d. None of the above

170. The abnormal post mortem pH values cause the defects like

a. PSE
b. DFD
c. Both of the above
d. None of the above

171. On the basis of muscle contraction speed, the muscles can be classified as

a. Slow and fast
b. Slow twitch and Fast twitch
c. Steady and speedy
d. None of the above

172. Name the carcinogenic hydrocarbons present in smoke

a. Benzo(α)pyrene
b. Polycyclic aromatic hydrocarbons (PAHs)
c. Both of the above
d. None of the above

173. Z-value represents °F required for thermal death time curve to pass through

a. One log cycle
b. Two log cycle
c. Three log cycle
d. None of the above

174. Cartilage of adult animal body is mainly present on

a. Disc of bone joints
b. Marrow of bones
c. Muscle surface
d. None of the above

175. Zinc is an essential component of_______enzyme

a. Heparin
b. Insulin
c. Rennin
d. None of the above

Q.2 Fill in the blanks

1. The total meat production of India is________________million tones.
2. Animal musculature is mostly of______________________in origin.
3. Meat is concentrated source of protein which are superior to the plant proteins due to________________________.

4. The homeostasis is controlled by nervous system which ceases within ____________ minutes after bleeding.
5. In wholesale cuts of buffalo carcasses, sides are separated and right side of carcass is called______________________.
6. In food preservation technique, irradiation destroys the micro-organisms by______________________.
7. In typical bacterial growth curve, bacteria multiply and increases in number in ____________ phase.
8. The presence of____________bacteria indicate faecal contamination in meat.
9. Pork fat becomes nearly ______________ in colour on cooking.
10. __________________ is the functional unit of myofibril.
11. Vitamin ______________ is almost absent in lean meat.
12. Formation of _______________ resulting in the onset of rigor mortis.
13. Presence of intramuscular fat is termed as ____________________.
14. Optimum humidity requirement for chilling of meat under refrigeration is __________ %.
15. Name the phase of growth of micro-organisms important for meat preservation ________________.
16. Fresh meat generally has water activity of_________________ or more.
17. MFPO stands for ____.
18. The chemical assay of________________is commonly used to determine amount of collagen in tissues.
19. The fatty acids of meat are relatively _______________particularly when compared to vegetable fats.
20. __________________is least firm among the meat from all food animals.
21. The water holding capacity of meat is__________________at isoelectric point.
22. _________________________is most commonly used flavour enhancer in meat products.
23. __________________is the most abundant protein in animal body.
24. ______________________accumulation causes lowering in muscle pH.
25. Manufacturer of meat food products who posses their own slaughter house are categorized in MFPO as______________________.

26. ______________________ mold is responsible for causing black spots in meat.
27. ______________________ resistant coated cans are primarily used for meat/ meat products.
28. Dehydrated meat should be packed in ______________________ films.
29. ______________________ present in smoke act as antioxidant.
30. There are ______________________ bacon factories in India at present.
31. India contributes ______________________ percent of total world beef meat trade.
32. The freezing point of muscle is about ______________________ ^{0}C.
33. Maximum permissible limit of nitrites in cured meat products is ______________________.
34. ______________________ act as chairman of Meat Food Products Advisory Committee.
35. Metmyoglobin is ______________________ in colour.
36. A-band of sarcomere is bisected by ______________________.
37. Actin and Myosin contribute about ______________________ percent.
38. ______________________ splits myosin into heavy meromyosin of length of actin filament.
39. ______________________ proteins are involved in regulation of length of actin filaments.
40. Diameter of skeletal muscle fibre ranges from ______________________.
41. Each actin filament is attached to ______________________ number of Z line filaments.
42. Each mammalian sarcomere has ______________________ number of triads.
43. ______________________ is most abundant protein in animal body.
44. The longitudinal tubules of sarcoplasmic reticulum converge in the H-zone of the sarcomere to form a perforated sheath known as ______________________.
45. Water holding capacity (WHC) of DFD meat is ______________________ than PSE meat.
46. Myoglobin ______________________ in colour.
47. ______________________ is the most important antibacterial agent present in smoke.

48. Beta actinin are involved in regulation of length of ____________.

49. ____________ meat products are most susceptible to light fading.

50. Heating of meat with nitrites produces ____________ which inhibits *Clostridium botulinum*.

51. ____________ muscle constitutes 35-65 % of the carcass weight of meat animals.

52. Structural unit of skeletal muscle tissue is called ____________.

53. FSSAI stands for ____________.

54. The distance between two adjacent Z-line is called ____________.

55. Stiffening in rigor mortis is due to formation of ____________.

56. Tenderization by electrical stimulation is mainly due to ____________.

57. In well bled muscle tissue, myoglobin constitutes ____________ of total pigment.

58. Collagen is a type of ____________ protein.

59. ____________ is an excellent method for preservation of meat.

60. Nitrate and Nitrite are commonly used compounds in curing but these compounds are also responsible for the formation of ____________ which is carcinogenic in nature.

61. Most preferred temperature for the growth of Psychrophiles is ____________.

62. Primal and sub primal cuts of fresh meat are susceptible with ____________ having excellent optical properties.

63. In patties preparation most desirable internal temperature is ____________ ^{0}C.

64. Name the media used for yeast and mould count ____________.

65. Name the bacteria responsible for green discoloration in meat ____________.

66. Requirement of electric current for stunning of poultry birds is ____________ volt A.C current for 1-3 sec.

67. Annual world meat production is ____________ million tones.

68. The share of bovine meat in total meat production of India is ____________ percent.

69. Extraction rate of cattle in India is ____________ percent.

70. Food Safety and Standard Bill was passed in the year _______________.
71. ___________________is the basic contractile unit of the muscle.
72. The word cattle is derived from Middle English word ___________________.
73. Meat course was first time started in the University of Minnesota by _________________.
74. The brown fat is mostly found around_______________in animals at birth then it starts disappearing.
75. In long bones hallow cylinder with long central shaft structure is known as____________.
76. The heritability of tenderness is approximately_________________percent.
77. Most preferred temperature for the growth of Psychrophiles is_____________.
78. The normal glycogen contents of mammalian muscle is ___________________ percent.
79. The normal glycogen contents in mammalian liver is _________________ percent.
80. Trichinella present in pork may be killed if the meat is heated to an ineternal temperature of ______________________________.
81. Muscle fiber diameter increases with_____________________of the animal.
82. F-value is the ____________________required to destroy organisms in specified medium at 250^0F.
83. Lethal rate is the reciprocal of_____________________________________.
84. For cold sterilization of meat, an irradiation dosages of________________ megarads is required.
85. The water exists in its three phases at a temperature of ______________and 4.7mm of Hg pressure.
86. Natural casings which are obtained from small intestines of animals, have lately been replaced by artificial casings made of_____________________.
87. The sheath of connective tissue surrounding muscle is known as ___________.
88. Cold shortening can be avoided by cooling the meat quickly to about _____________and holding at it for the onset of rigor.
89. The zone of maximum ice-crystal formation during freezing of meat ranges from_____________.
90. The human taste buds can basically characterize food into Sweat, Salty, Sour and _________________taste categories.

91. Saran which has excellent barrier and ______________________ ______.

92. Formation of enterotoxin by *Staphylococcus aureus* is occasionally encountered in fresh meat due to rapid proliferation of ______________ ____________________.

93. Corresponding muscles of pigs have ______________________ tissue when compared to cattle.

94. In horse meat cytochrome oxidase activity is ______________ in younger animals of 2 years when compared to older animals of 4 years.

95. Glycogen level of newly born pork is high (around 7%) because ________.

96. Identification of species of fresh fish can be done based on their ________________ through electrophoresis technique.

97. Drip losses in lamb meat increases rapidly during thawing when shortening exceeds a critical level of ____________________.

98. Inosine -5-monophosphate in meat________________during post rigor stage when potential meat flavour is developing.

99. In comparison to cattle, buffalo meat is darker due to ______________.

100. Susceptibility of cured meat pigment to irradiation by fluorescent light is__________at higher pH.

101. Excepting ________________most of the pathogenic bacteria do not grow below a_W 0.91.

102. The red muscle has a limited ________________ activity and high content of mitochondria.

103. Myofibril account for________________ of muscle cell volume.

104. The light bands of muscle are bisected by __________________disc.

105. The_____________protein is closely associated with tails of myosin molecules.

106. Maximum reversible contraction varies from______________________.

107. ________________ acts as a plasticizer allowing separation of actin and myosin.

108. ________________strength of collagen increases with ageing.

109. The characteristic colour of meat is a function of meat____________ and scattering properties.

110. Meat from male animals contain______________myoglobin as compared to females of the same age group.

111. Myoglobin in fresh meat, exists in three forms and has a great affinity for_______________.

112. Characteristic flavour of meat develops during_______________________.

113. During freezing storage of pre-rigor meat at_____________ or lower, there is little change in ATP.

114. The fine streaks of fat visible in pig muscles between the ribs is called _____________.

115. Connective tissue proper consists of structure less mass called________.

116. The critical level of Ca++ ions concentration is necessary to initiate contraction and relaxation of muscle is__________________.

117. A_______________ fold increase in Ca++ ions concentration is necessary to initiate contraction.

118. Maintenance of physiologically balanced internal environment is known as_________________.

119. There is______________in muscle temperature soon after exsanguinations.

120. Common degree of thaw rigor shortening is _____________ of its original length.

121. Attaining of pre-rigor temperature between 0 and 15^0 causes ___________.

122. Lack of space for water molecules within protein structure due to presence of bivalent ions is called______________________.

123. Cathepsin the proteolytic enzymes are held in organelles called ________.

124. Muscles with high marbling tend to have__________________ WHC.

125. During conditioning, the temperature is controlled to a maximum of _________________.

126. A combined treatment of high pressure and heat treatment to______________ improves tenderness in beef.

127. Ionizing radiation at doses of Ca_____________ has been used to eliminate Salmonella.

128. Hotboning involves removal of flesh in pre-rigor condition upto 3 hrs after slaughter when muscles are still warm at ______________________.

129. Electrical stimulation is used during processing of beef and lambs but not pork in due to significant risk of ___________________________.

130. Maximum tenderness of meat is associated with an intermediate rate of glycolysis and not the very____________________ rate obtained using electrical stimulation.

131. Cold shortening is a ____________________process.
132. Electrical stimulation is used to prevent cold shortening by ______________ post mortem glycolysis.
133. Basically mechanical deboners can be classified into _____________ different groups.
134. The recommended thawing procedure is to thaw meat products at _________ temperature.
135. Storage temperature of ______________ are recommended for both commercial and home freezer units.
136. Cryogenic freezing can be carried out by any of immersion, spray, circulation near product systems and the commonly used cryogenic agents are ___________________.
137. The cold shortening of muscle is more rapid at 10-14^{0}C and minimum at ____________.
138. The water activity of fresh meat is generally _____________or above.
139. The optimum water activity for the growth of most bacteria is in the range of_________.
140. The minimum water activity for Salmonella has been reported near________________.
141. The minimum water activity for Staphylococci is near__________________.
142. The time necessary to completely sterilize a suspension of bacterial cells or spore at a given temperature is known as_______________________.
143. *Tenia solium* is found in the intramuscular connective tissues of__________.
144. *Tenia saginata* is found in the intramuscular connective tissues of________.
145. The most suitable machine to test the tenderness of meat with auto recording system is_________________.
146. Frozen buffalo, sheep and goat meat accounts for____________ of all annual export of India.
147. India exports annually casings of worth Rs____________ crores.
148. Major drawback of use of biotechnological manipulation in meat animals is ____________________.
149. _________________ would be the priority area for growth and development of meat industry in India.
150. Myoglobin content is high in ____________________.

151. ______________________ meat is rich in cholesterol.

152. The musculature of sheep, cattle and pigs consist of about________________ anatomically distinct units.

153. Reticulin and collagen can be differentiated by ____________________ staining.

154. ______________________ are responsible for the mechanical strength of collagen.

155. At high ultimate pH the degree of proteolysis is________________________.

156. The most important manifestation of postmortem denaturation of the muscle protein is____________________________.

157. The rigor reflect the abnormal post-mortem stimulation of________________.

158. State of iron in metmyoglobin is______________________________.

159. Electrical stimulation of the lamb carcass should be done within __________.

160. Skeletal muscles constitute about _____________________ of the carcass weight.

161. Skeletal and __________________ muscle are also referred to as striated muscle.

162. Smooth and cardiac muscles are called _____________________ muscles.

163. There are more than _________________ muscles in animal body.

164. The structural unit of skeletal muscle tissue is referred as________________.

165. Membrane surrounding muscle fibre is called________________________.

166. Troponin and Tropomyosin are called ____________________________.

167. Myosin constitutes approx. ________________ of the myofibrillar proteins.

168. The adipose tissue contained in connective tissue septa between the individual muscles is known as______________________fat.

169. Ratio of dressed carcass weight to weight of live animal known as_________________.

170. Accumulation of lactic acid as a result of anaerobic glycolysis in muscle results in lowering of _________________.

171. There is _____________ in muscle temperature soon after exsanguinations.

172. Onset phase of rigor is preceded by______________phase.

173. Common degree of thaw rigor shortening is_____________ of its original length.

174. Attaining of pre-rigor temperature between 0 and 15^{0}C causes ____________.

175. Myoglobin constitutes ________________ of total meat pigment.

176. Exchange of ions in muscle proteins during ageing results in improved ____________________.

177. Resolution of rigor mortis is a result of alterations in structure of myofilaments including ____________________.

178. When carcasses are suspended from Achilles tendon the ____________ muscle becomes extremely tender.

179. When carcasses are suspended from the pelvic (aitch) bone muscles around ______________ become more tender.

180. There is no difference in the tenderness of the muscle ______________ when suspended either from Achilles tendon or pelvic bone.

181. Relative humidity requirement for growth of moulds on the meat surface is______________.

182. Extensive lypolysis caused by microbial agents bring about a ____________ flavour.

183. Blood pH ____________ drops in stress resistant pigs.

184. Water holding capacity of meat is______________ at ultimate pH.

185. Sarcomere length is ________________ at rigor.

186. At rigor____________ of cross bridging sites are engaged.

187. Delay phase is followed by ___________ phase in development of rigor mortis.

188. Molecular weight of myoglobin is ___________times that of haemoglobin.

189. Each actin filament is surrounded by ____________myofilaments

190. Each myofilament is surrounded by ____________ filaments.

191. In muscles of meat animals contain ___________proportion of white fibres than red fibres , even in muscles of visible red colour.

192. Bright red colour of meat is due to_________________.

193. Ultimate pH of meat is about 5.5 which is closer to the ______________ of the principal proteins in muscles.

194. In fresh meat, before cooking the most desirable form of myoglobin is_____________.

195. The principal colour pigment in case of cooked bacon is ____________.

196. A chemical ______________________ is responsible for typical boar odour.

197. The carcass yield of pig is around __________ and that of sheep is around_________.

198. Fish was first canned in ____________________.

199. ____________________ is called the father of canning.

200. The pink or reddish liquid that comes from meat on thawing is called______________.

201. When ice crystals evaporate from an area at its surface a defect is called____________.

202. The use of SO_2 as a meat preservative originated in _________________.

203. Sorbic acid was approved as food preservative in _________________.

204. Antibiotic___________________ approved for use in fresh poultry in 1955.

205. ____________________was the first chemical preservative permitted in foods by the US FDA.

206. Acetic acid and lactic acid bacteria and many molds are more sensitive to SO_2 than_______________.

207. Propionates are _______________rather than fungicidal.

208. Sorbic acid is ineffective at about pH________________.

209. Plasmolysis is nothing but________________ of cells.

210. ______________ is a unit of measure used for expressing exposure does of gamma radiation.

211. _______________ is equivalent to pasteurization.

212. The shelf life of shrimp is extended from _______________ fold by radurization.

213. ________________freezing is the process by which temperature of food is lowered to about-20^{0}C within 30 minutes.

214. __________________ drying precedes constant rate drying.

215. _______________ refers to degrees F required for thermal destruction curve to traverse are log cycle.

216. A radiation dosage upto________________has been approved by WHO as being "unconditionally" safe for human consumption.

217. The genera Bacillus and Clostridium contain the ______________ of greatest importance in foods.

218. Quick or fast freezing is the process by which the temperature of foods is lowered to about -20 ^{0}C within______________minutes.

219. *Micrococcus radiodurans* is one of the most______________________ bacteria known.

220. Antimicrobial substance present in garlic is known as ______________.

221. IMF are characterized by a moisture content of around 15 to 50 and a_w between_________________.

222. The antibiotic _________________is a polyene that is quite effective against yeasts and molds but not bacteria.

223. An elastic protein, connectin contained _________percent lipid and one percent carbohydrate.

224. ______________ of proteins has been cited as the main factor responsible for the formation of good quality gel.

225. In addition to actin and myosin _________________ constitute filaments cytoskeleton.

226. Sarcoplasmic reticulum constitutes approximately about __________ percent of total fiber volume.

227. T-tubules constitutes approximately about _____________percent. of total fiber volume.

228. When colloidal dispersion of some large molecules are cooled, the viscosity increases to a point at which some rigidity is attained, which is called as___________________.

229. The odour threshold of hydrogen sulphide in water is ________________.

230. Freshly cooked chicken meat has _______________times more hydrogen sulphide than the threshold value.

231. Fatty acids on oxidation produce _____________compounds that are potential flavour contributors.

232. Many gels on continued standing loose solvent and gels shrink in a process called______________________.

233. Anaerobic glycolysis of glucose yields______________ molecules of ATP.

234. In longissimus dorsi muscle, the muscle fibers are arranged______________.

235. In crossed section of the fixed muscle the myofibrils appears to be subdivided into groups which are well known as__________________________.

236. Among the muscle proteins ______________________ fraction is mainly responsible for the formation of good quality gel.

237. ______________________ are associated with sarcoplasmic reticulum.

238. Myosin constitutes 55-60% of myofibrillar proteins in muscle with a molecular weight of ______________________.

239. In resting muscle, the dark 'A' band extends along the axis of contraction and about______________long.

240. The actin content of the muscle is______________________.

241. Red muscle fibers contain high proportion of enzymes involved in ____________.

242. There are about ______________ myosin molecules for a complete thick filament.

243. ____________________is the most abundant fatty acid in the animal body.

244. Two steps involved in destabilization process of an emulsion are floculation and ______________________.

245. __________________is identified as the substance responsible for unpleasant boar odour in pork on heating.

246. Oxidation of methionine results in ________________, a component of meat flavour.

247. Each muscle cell is surrounded by a collagenous reticular layer called____________.

248. Among all the cytoskeletal proteins_____________ is the most prevalent, accounting for 6% of the myofibrillar bulk.

249. The main protein of thin filament is_________________.

250. In the presence of initiaters such as heat, metals or light, unsaturated lipids (LH) lose a hydrogen radical to form ____________________.

251. IMP in meat is derived from adenosine-5-triphosphate (ATP) via adenosine-5-monophosphate (AMP) through the action of enzyme__________________.

252. Reticulin exihibits ________________ staining with amonical silver contrast staining.

253. Collagen exihibits________________staining with amonical silver contrast staining.

254. At tropocollagen the fibres are laid down in____________ lattice which oriented at an angle to the long axis of muscle.

255. At the amine end of each tropocollagen the chains are__________________ helical.

256. The principal mucopolysaccharides of the ground substance of muscle is _________.

257. The distance between two adjacent Z-line is functional unit of myofibril known as__________________.

258. _____________ myosin filament is surrounded by six actin filaments in hexagonal array.

259. There are ________________light subunits found in the head reason of myosin.

260. Mitochondria of skeletal muscle are particles situated at ______________ and are specially prevalent in active muscle.

261. The _______________ tubules transmit the excitatory impulse in contradiction from depolorized sarcolemma.

262. G-actin polymerizes into F-actin in the presence of _________________.

263. The characteristics yellow colour of elastin is due to _________________ residue. (tetramethyl substituted pyridine)

264. Elastin differ from collagen in having only_____________ percent hydroxyproline.

265. Haemoglobin and myoglobin are purplish red whereas, oxymyoglobin and oxyhaemoglobin are ________________red in colour.

266. The green pigment are formed in meat when alpha_______________bridge is attacked.

267. Fatty acid _________________leads to the formation of carbonyl compounds which are important flavour contributors.

268. In sausage emulsion water forms the continuous phase and _______________ forms discontinuous phase.

269. The structural unit of collagen fibril is called______________________.

270. The most immediate source of energy that can be mobilized for ATP sysnthesis is ____________________.

271. The weight of molecular tissue ranged from __________________ percent in sheep.

272. Discolouration in fresh meat is predominantly caused by oxydation of the meat pigments to______________________brown pigment).

273. To prevent desiccation in fresh meat, a relative humidity of ____________% is used.

274. The ________________ are among the principal mucopolysaccharides of ground substance.

275. Mechanically deboned meat has ________________ calcium content as compared to hand deboned one.

276. At level of ________________ of meat, off odours and possibly slime may be observed.

277. Salmonella should not be detected in more than ____________ of the five 25 g samples.

278. Black spots on meat are a result of a mold ________________________.

279. White spots on meat are caused by a mold __________________________.

280. Dark cutting meat condition is common in stress ____________________ animals.

281. ___________________________gene is responsible for acid meat.

282. The typical brown colour of smoked meat is due to ____________________.

283. Relative humidity of chilling should be within the range of ______________.

284. Typical colour of beef is ____________________________.

285. ____________________contributes to the mechanical stability of muscle filament.

286. Meat is generally good source of all minerals except__________________.

287. Among all food animals, meat of __________________ is least firm.

288. Apparent resolution of rigor mortis is due to degradation of structure at_____________.

289. Food Safety and Standard Bill was passed in the year___________________.

290. The rough shoulder in pork carcass is separated from rough posterior middle and rough ham by a cut between _________________ perpendicular to spine.

Q.3. Write true or false against each statement

1.	HACCP is a comprehensive food safety system.	( )
2.	Flavour is a complex sensation of taste and odour.	( )
3.	Specialized panel testing usually precedes consumer panel.	( )
4.	Double Immuno-diffusion test was developed by Ouchterlony.	( )
5.	ELISA is an important qualitative immunological tool for meat species speciation which is not monitored by precipitation.	( )
6.	Quick air blast freezing is widely used for long term storage in commercial enterprises.	( )
7.	Extract release volume (ERV) of meat increases on spoilage.	( )
8.	Irradiation resistance of bacteria is highest in its lag phase.	( )
9.	Aseptic packaging is suitable for liquid products only.	( )
10.	Meat is a good source of calcium.	( )
11.	H-zone has only myosin filament.	( )
12.	Fresh meat should be packed in water and oxygen im15permeable film.	( )
13.	Oxymyoglobin is most commonly occurring oxidized form of myoglobin.	( )
14.	Consumer panel is involved in sensory evaluation before specialized panel.	( )
15.	Cellophane is an environment-friendly packaging film.	( )
16.	Fresh beef meat is always packed in oxygen permeable film.	( )
17.	Meat is good source of Vitamin-C.	( )
18.	Surface slime of meat is caused by Anaerobic bacteria.	( )
19.	There are 3600 licensed slaughter house in India.	( )
20.	Irradiation resistance of micro-organisms is higher in the presence of oxygen.	( )
21.	Canned meat are usually spoiled by Anaerobic bacteria.	( )
22.	Ultimate pH of meat is high in DFD meat.	( )
23.	Cardiac muscles are involuntary in nature.	( )
24.	Average length of resting mammalians muscle sarcomere is about 25 millimicron.	( )
25.	Tails of myosin molecules are functionally active sites of thick filaments during muscle contraction.	( )
26	Saturated fatty acids are more resistant to oxidative rancidity than unsaturated fatty acids.	( )
27	Micro-organisms which can grow with or without presence of oxygen are known as halophilic organism.	( )
28	Smooth muscles are uninucleated.	( )

29	About 95 percent of the total water in muscle exists as free water.	( )
30	Isoelectric point of myosin is 4.7.	( )
31	Fasciuli of muscle are covered by connective tissue sheath is known as andomysium.	( )
32	Slow freezing favours the formation of large ice-crystals.	( )
33	Meat Food Products Order (MFPO) were enforced in 1975.	( )
34	Livestock products contribute about 35.05 percent of total GDP.	( )
35	KCl has lower ionic strength than NACl.	( )
36	Nuclei of mammalian muscles are ellipsoidal in shape and centrally located.	( )
37	The connective tissue sheath surrounding muscle is known as endomysium.	( )
38	*Sporotrichum cornis* is the most common cause of white spot in meat.	( )
39	Potessium sorbate is most commonly used flavour enhancer in meat products.	( )
40	Glycine is the most abundant amino acids in collagen.	( )
41	Each actin filament is connected to eight Z-filaments that pass obliquesly through the Z-line and connects to different actin filaments in adjascent sarcomere.	( )
42	PSE condition is most commonly observed in pork.	( )
43	Water holding capacity of muscle tissue has direct effect on shrinkage of meat during storage.	( )
44	Marbling contributes to firmness of refrigerated meat.	( )
45	The mode of action of red muscle fiber is tonic in nature.	( )
46	Actin is a major myofibril protein contributes about 5% in the total myofibril proteins.	( )
47	Tropomyosin is a highly charged protein molecules and fibrous in nature.	( )
48	Napole gene is responsible for DFD meat.	( )
49	Loss of homeostasis occurs within 4-6 hrs of exsanguinations of animals.	( )
50	pH is an important factor in determining the water holding capacity.	( )
51	BIS standards are involuntary in nature.	( )
52	India is the third largest consumer of pesticides in the world.	( )
53	The average dressing percentage of poultry birds is 50%.	( )
54	US standards and grade for dressed poultry were established in the year 1927.	( )

55	In grading characteristics of chicken if freezer burn is in the area of 0.5 cm in diameter then it will come under Grade -1.	( )
56	U V rays have strong penetration power.	( )
57	Tandoor is smokeless moderate and niform heating device generally maintained at 500⁰F.	( )
58	Salmonella is a most uncommon egg borne disease in human beings.	( )
59	Red muscle fibers are more vulnerable to cold shortening than white muscle fibers.	( )
60	Rigor mortis is completed early in PSE meat.	( )
61	Organic acids in smoke have major preservative effect.	( )
62	Loin is the upper middle portion of pork cut.	( )
63	Malaysia is the major importer of Indian meat.	( )
64	Sheep was the first animal domesticated in East Asia about 15000 years ago.	( )
65	The mode of action of red muscle fibre is tonic in nature.	( )
66	Actin is a major myofibril protein contributes about 5% in the total myofibril proteins.	( )
67	Tropomyosin is a highly charged protein molecules and fibrous in nature.	( )
68	Napole gene is responsible for DFD meat.	( )
69	Loss of homeostasis occurs within 4-6 hrs of exsanguinations of animals.	( )
70	The pH is an important factor in determining the water holding capacity.	( )
71	BIS standards are involuntary in nature.	( )
72	India is the third largest consumer of pesticides in the world.	( )
73	White muscle fibres are generally larger in diameter than red fibres.	( )
74	Haemoglobin has higher affinity for oxygen than myoglobin.	( )
75	G-actin has higher molecular weight than F-actin.	( )
76	Tropomyosin-B is a highly unique protein containing 80-100 percent alpha helical structure in normal solvents.	( )
77	Lower the temperature of colour, greater will be the salt penetration during curing of meat.	( )
78	Spices used in meat products serve as bacteriostatic agents.	( )
79	Myofibrillar proteins decreases with activity of the animal.	( )
80	The meat from old animals shows coarse marbling.	( )
81	Decimal reduction time is the time of heating at a temperature to cause 10 percent reduction in the count of viable cells or spores.	( )

82	Collagen content of more than 25 percent is desirable for good quality sausages.	( )
83	Bioriented polyethylene film is used for shrink packaging of meat.	( )
84	Meat proteins at a pH lower than their isoelectric point exhibit higher water binding capacity.	( )
85	Pre-slaughter administration of calcium salts enhances drip losses during thawing of frozen meat.	( )
86	The attractive red colour of cured meat is essentially due to the presence of nitrosohaemochromogen.	( )
87	A minimum of 100 ppm nitrite is required to inhibit *Clostridium botulinum.*	( )
88	If a more strongly the ions are bound by meat proteins, the weaker will be their hydrating effect.	( )
89	Carcasses should be kept around 35-40^{0}C during curing.	( )
90	Enzymes and viruses are more resistant to irradiation than bacterial spores.	( )
91	Freeze drying is better than canning and irradiation in maintaining the vitamin content of meat.	( )
92	Addition of sodium chloride improves WHC of meat.	( )
93	Refreezing and thawing improves quality of meat.	( )
94	High ultimate pH causes pale colouration of meat.	( )
95	Ultimate pH of post mortem muscle is 5.8.	( )
96	Calcium content of mechanically deboned meat is higher than that of normal meat.	( )
97	Collagen fibrils do not branch but fibers do.	( )
98	The main function of M-line is to keep A-filaments in position in the longitudinal as well as transverse direction.	( )
99	In smoked meat the onset of rancidity is enhanced.	( )
100	The nitrate is reduced to nitrite by nonpathogenic organisms in meat.	( )
101	The sulfur compound contributes poorly to meat flavour and odour.	( )
102	Sodium chloride increases the water holding capacity and swelling of meat on the acidic side of the iso-electric point.	( )
103	Preservation of meat by irradiation is widely accepted.	( )
104	*Staphylococcus aureus* is very tolerant to salt and is not inhibited by any permissible combination of curing salt.	( )
105	Tenderness of meat increases with increase in age.	( )
106	During cooking the muscle fibre becomes tender and the connective tissue tough.	( )

107	Frozen buffalo meat is the single largest item imported by Malayasia from India.	( )
108	Rabbit meat is lean meat and its palatability can be rated above than pork and mutton.	( )
109	Rock salt is most suitable for salting and curing of casings.	( )
110	Evaporative weight loss during chilling can be avoided by hot boning and vacuum packaging.	( )
111	Trimming of fat to a consistent standard may be difficult with hot meat.	( )
112	Electrical stimulation with hot boning has no obvious advantages in tenderizing meat other than the prevention of cold shortening effect.	( )
113	Cold shortening is the common phenomenon in improperly preserved fresh meat cuts.	( )
114	Conditioning/ ageing is routinely done in meat processing plants in view of the advantages.	( )
115	Heat tenderizing/ enzymes gets activated during heating.	( )
116	Castrated cattle have more intramuscular fat than corresponding uncastrated animals.	( )
117	Connective tissues content of muscle is greater in young animals than older ones.	( )
118	Final pH (ultimate pH) attained is always the result of lack of glycogen in muscle.	( )
119	In rigor mortis all the available binding sites are utilized to form actomyosin cross bridge as against only 20% in routine contraction.	( )
120	Frozen meat attracts fewer consumers in India due to quality deterioration during freezing.	( )
121	Most of the animals slaughtered in India produce DFD meat.	( )
122	High pressure tenderization is more beneficial in pre-rigor meat.	( )
123	Meats preserved by irradiation are superior in quality to thermally processed meats (canned product).	( )
124	Major effect of high pressure tenderization in meat is depolymerization of myofibrillar proteins.	( )
125	Fresh meat in different types of shortening can be effectively assessed by consumer palate.	( )
126	Transgenesis is a common practice adopted in rearing / production of meat animals.	( )
127	Implantation of anabolic agents is associated with the decrease in tenderness , juiciness and flavour of meat.	( )
128	DFD meat is safe to consume.	( )

129	Nutritive value of doppalander meat is as good as normal meat.	( )
130	Safe dose of ionizing radiation applicable in meat foods and recommended by WHO is 10Gy.	( )
131	Growth curve in meat animals is always 'S' shaped.	( )
132	Irradiation of all types of meat foods is permitted by Govt. of India.	( )
133	Myoglobin is myofibrillar protein.	( )
134	Troponins are sarcoplasmic proteins.	( )
135	Muscle is free from lactic acid.	( )
136	Actin content is more than myosin in meat.	( )
137	Pre-rigor meat has better water binding capacity.	( )
138	The thin filament of actin are continuous through the Z-line but do not transverse the A-band only.	( )
139	Myosin filament transverse the A-band only.	( )
140	pH fall is not affected by holding temperature of meat.	( )
141	Injection of calcium salts enhance the acceleration rate of post-mortem glycolysis.	( )
142	Disappearance of ATP from the muscle is responsible for the formation of actomyosin complex.	( )
143	Acid rigor is common in immobilized animals.	( )
144	Rapid pH fall prevents cold shortening.	( )
145	The higher ultimate pH, the better is the keeping quality of meat.	( )
146	Dissolution of actin-myosin complex occurs during conditioning of meat.	( )
147	Meat freezes at 0^0C.	( )
148	Oxidative rancidity is retarded by high ultimate pH.	( )
149	Conversion of muscle to meat is a aseptic process.	( )
150	Meat is consumed for vitamins and minerals.	( )
151	Gap filaments have major role in contraction of muscles.	( )
152	Addition of sodium chloride improves WHC of meat.	( )
153	Refreezing improves the quality of meat.	( )
154	Shrink packaging is generally used for uneven meat cuts.	( )
155	High ultimate pH cause pale colouration of meat.	( )
156	The heads of meromyosin are globular in nature and are about $200A^0$ long.	( )
157	Actin- to- myosin center –to- center, distance increases on contraction and decreases on stretch , in a proportion reciprocal to the square root of the muscle length.	( )

158	In mammalian muscle tissue transverse elements of the sarcoplasmic reticulum occurs at A-I junction but in fish they occur at Z-line.	( )
159	Muscle contraction begin when the Ca++ concentration increases to about 10^{-7} M.	( )
160	In lamb meat cuts the ratio of elastin-to-collagen is less than 0.1, while in beef it is 0.2.	( )
161	During postmortem, muscle tissue becomes rigid like rigor mortis, in the absence of ATP due to cross linking between actin and myosin filaments.	( )
162	Rapid loss of extension begins in horse muscle when ATP level falls to 50% of its original value.	( )
163	Shortening of unrestrained excised sternomandibularis muscle obtained from beef is more at 2^0C when compared to 19^0C.	( )
164	The total adenosine triphosphate activity is lower in thaw rigor muscle tissue than in normal rigor muscle tissue, at the same temperature.	( )
165	The muscle with a long sarcomere length and small fiber diameter is usually less tender than its counterparts.	( )
166	The recommended procedure is to thaw frozen meat is in water.	( )
167	Bone darkening is quite common in young chicken/ fryers.	( )
168	The effect of temperature fluctuation during frozen storage of meat is most damaging.	( )
169	Bone darkening does not affect the flavour, odour or texture of cooked meat.	( )
170	Achilles tendon hanging brings about tenderness in more number of muscles as compared to Pelvic hanging.	( )
171	Thawing and refreezing of meat is advisable.	( )
172	Time temperature pattern of thawing is more detrimental to quality than that of freezing.	( )
173	Most chemical changes could be eliminated by reducing temperature to -40^0C.	( )
174	Lean to fat ratio influences the freezing rates of meat products.	( )
175	The haemoglobin molecule has molecular weight four times that of myoglobin molecule.	( )
176	Pink or green discolouartions are sometimes encountered in the fat of cured meat caused due to excessive use of nitrite.	( )
177	Ageing or conditioning of meat increase its water holding capacity.	( )

178	Shortening in unrestrained muscles is more during cold shortening than during thaw rigor.	( )
179	Cooking makes connective tissue more tender by converting collagen and elastin to gelatin.	( )
180	Cooked pork contains greater quantity of H_2S than cooked beef because of its higher level of free SH group.	( )
181	DFD condition is observed in pork having low ultimate pH.	( )
182	Low freezing temperature of about -20^{0}C are less harmful to micro-organisms than-10^{0}C.	( )
183	Fat decreases the heat resistance of micro-organism.	( )
184	Drying is same as dehydration.	( )
185	Low temperature preservation of foods is known as cold sterilization.	( )
186	Bone darkening is prevalent more in young chickens.	( )
187	Case hardening is a problem during freeze drying of meat.	( )
188	Fast freezing results in greater drip loss.	( )
189	TDT is same as Z-value.	( )
190	PA3679 is found in frozen foods.	( )
191	Cans must not be cooled to temperature below 35^{0}C.	( )
192	Freeze dried meat must be vacuum/inert gas packed.	( )
193	Percent reconstitution=DR/RRx100	( )
194	Medium acid foods have pH between a 5 and 4.5.	( )
195	Peroxidase is the most heat resistant to enzyme.	( )
196	Since Clostridium botulinum spores germinate and produce toxin below pH4.6, 12-D concept is observed only for foods below this pH value.	( )
197	Nitrates in meat helps to kill spores of anaerobic bacteria by heat and inhibit germination of spores.	( )
198	Enterococci survive freezing and persist longer during storage than coliform bacteria and therefore, are recommended as 'indicator bacteria' for possible faecal conatamination in frozen foods.	( )
199	The F-value is a decisive factor for the storage life of the canned product.	( )
100	The higher the optimal and maximal temperature for growth, the lesser the resistance to heat is likely to be.	( )
201	Bacterial cells show their greatest resistance to heat during their phase of logarithmic growth.	( )
202	In microbial growth curve phase of positive acceleration is followed by the phase of negative acceleration.	( )

203	Most organisms are more tolerant of low a_W at pH values near neutrality than in acid or alkaline media.	( )
204	The inhibitory effects of salt (NaCl) are dependent upon pH as are some other chemical preservatives.	( )
205	Gram negative bacteria are more resistant to radiation than gram positive bacteria.	( )
206	Sterilization doses of radiation are insufficient to destroy the natural enzymes in food.	( )
207	Endospores of spores forming bacteria and toxins of food poisoning bacteria are apparently unaffected by low temperature.	( )
208	More number of metabolically injured bacterial cells can be recovered on highly nutritive non selective media than on selective media.	( )
209	Freezing has no effect on cellular proteins of micro-organism.	( )
210	Intermediate moisture foods (IMF) produced by desorption is more inhibitory to micro-organisms than that produced by adsorption.	( )
211	Emulsifying agent functions to reduce the interfacial tension between the two phases. Thus reducing the energy input in the formation of emulsion and increases its overall stability.	( )
212	Fatty acids occurring naturally in animal fat are mostly in cis form.	( )
213	Intact myofibrillar and sarcoplasmic proteins as such do not contribute to meat flavour.	( )
214	Sarcoplasmic reticulum is a system of vesicles and tubules that surrounds each muscle fiber.	( )
215	Both thick and thin filaments are present in 'I' band region of the myofibril.	( )
216	Each actin filament connects to four Z-filaments near Z-line.	( )
217	The off flavour referred to as warmed over flavour in cooked, stored and subsequently warmed meat is due to oxidation of short chain fatty acids existing in muscles.	( )
218	Sarcoplasmic reticulum is not well developed and terminal cisternae are absent in smooth muscles.	( )
219	Sarcoplasmic and stromal fractions of proteins posses good gel forming ability.	( )
220	Gel forming potential of myosin is located in myosin head.	( )
221	Light meromyosin carries the biological activity of myosin.	( )
222	Disodium glutamate is a potent flavour enhancer as monosodium glutamate and inosinate.	( )

223	BHT owes its antioxidant property due to the hindered phenol structure.	( )
224	Lipid oxidation is a major cause of deterioration in the quality of meat products stored at low temperature.	( )
225	ATP is more important than Ca++ ions for muscle contraction to start.	( )
226	The keto amino acids decompose to give carbonyl intermediates in browning reaction.	( )
227	Phosphates are effective antioxidants.	( )
228	Metmyoglobin is the oxidized form of myoglobin and cannot bind oxygen.	( )
229	Higher the HLB value more hydrophilic the emulsion is favouring oil in water type emulsion.	( )
230	The collagen fibers are randomly arranged as a criss cross lattice which confers necessary degree of flexibility and extensibility on skin.	( )
231	Higher ultimate pH lowers the flavour intensity in meat.	( )
232	DTNB light chain of myosin has ATPase activity.	( )
233	Tropomyosin has a very high charge density because of its 19% basic and 25% acidic amino acid residues.	( )
234	Glycolysis is the chief source of energy for resphosphorylation in white muscles.	( )
235	Red muscles generally use fatty acids as their major fuel and white muscles, on the other hand, use glucose as major fuel.	( )
236	Anserine and cornosine are the natural flavour compounds developed on cooking of meat.	( )
237	More of the short chain fatty acids and triglycerides are emulsified readily than the long chain fatty acids.	( )
238	Myoglobin molecule is four times as large as the haemoglobin.	( )
239	High temperature will promote coalescence of oil droplets causing less oil to be emulsified before the capacity of proteins to emulsify is exceeded.	( )
240	The falvour components identified from pork and lamb lean are identical to those in beef.	( )
241	The colour of myoglobin pigment is bright red.	( )
242	Metmyoglobin is formed by combination of metmyoglobin with nitric oxide.	( )
243	Myoglobin is one of the least heat stable of sarcoplasmic proteins.	( )
244	If the ultimate pH of meat is high the surviving activity of the cytochrome enzyme will be greater.	( )

245	Species specific flavours can be related to subtle differences in the amino acid and carbohydrate content of adipose tissue.	( )
246	At very low levels, water has an antioxidant effect.	( )
247	The secondary products identified in highly auto-oxidized eleate include allylic ketones.	( )
248	The titre of lenoleic acid is greater in the lean meat of sheep than in that of pigs.	( )
249	Unlike collagen and reticulin , elastin is net broken down on exposure to heat.	( )
250	The hydroxyproline content of muscle is used as a measure of its adipose tissue.	( )
251	Intermolecular cross linkage is not responsible for strength of collagen.	( )
252	The schiffbase linkage of collagen neither heat nor acid labile.	( )
253	With maturation of the animal the cross links are stabilized in non-reducible forms.	( )
254	Myosin has poor affinity for calcium and magnesium.	( )
255	Myosin molecules are built of two subunits L and H meromyosins.	( )
256	Tropomyosin extends along the helical groove in actin filament.	( )
257	B-actin inhibits the polymerization of F-actin.	( )
258	The flavour precursors of meat are water insoluble.	( )
259	Non-enzymatic browning reaction between amino acid and reducing sugar in meat gives aroma.	( )
260	Collagen is not converted to water soluble gelatin at 80^0C.	( )
261	The contractile ATPase activate the contraction.	( )
262	Cathpsin is found in mitochondria of muscle cells.	( )
263	The molecular weight of smooth muscle myosin is greater than that of striated muscle.	( )
264	The major calcium binding protein in smooth muscles are Calmodulins.	( )
265	The major calcium binding protein in the striated muscles is Troponin-C.	( )
266	Hydroxyproline is rarely found in proteins other than collagen and elastin.	( )
267	Collagen is most abundant protein in the body of the animals.	( )
268	The proportion of salt soluble and acid soluble collagen is greater in tissues from young than from old animals.	( )
269	TCA cycle enzymes are found in microsomal fractions of sarcoplasmic proteins.	( )

270	Myosin is resistant to the proteolytic action of trypsin.	( )
271	Skeletal muscles have the higher amount of cholesterol than smooth muscles.	( )
272	Meat emulsion is water in oil type.	( )
273	The emulsifying capacity of salt soluble proteins is maximum at pH 6.0-6.5.	( )
274	Lysine is the most sensitive amino acid to maillard reactions.	( )
275	Increasing pH and processing temperature are the easiest ways of producing LAL formation in muscle foods.	( )
276	Alpha actinin accelerates polymerization of G-actin to F-actin.	( )
277	Actomyosin is the major form of myofibrillar proteins that is found in post mortem muscle.	( )
278	G-actin polymerizes in F-actin in presence of salt and small amount of ATP.	( )
279	In addition to amino acids, carbohydrates and fats, thiamin is an important precursor of meat aroma.	( )
280	Dark colour of pre-rigor muscle is due to oxygen consumption by actively respiring mitochondria at muscle surfaces keeping myoglobin deoxygenated.	( )
281	The meat flavour of fresh meat increases with conditioning.	( )
282	Oxidized flavour which develops rapidly in refrigerated cooked meats is termed WOF.	( )
283	Unsaturated fatty acids in meat fats with triple bonds are not found.	( )
284	Oxidative changes are more in tissues containing phospholipids than in those containing neutral lipids.	( )
285	Degradation of glucose-l-phosphate anaerobically resulting into post mortem accumulation of lactic acid by the Embdenmeyerhof system is called glycolysis.	( )
286	The rate of oxidation of intramuscular fat will be higher in meatfrom animals on a low plan of nutrition.	( )
287	Metmyoglobin is the most commonly occurring undesirable pigment on meat surfaces.	( )
288	Type-I collagen is found in muscles and tendons.	( )
289	Reticulin does not yield gelatin on boiling.	( )
290	A narrow dark line in the middle of the I-band is called Z-band.	( )
291	The sarcomere is the repeating structural unit of the myofibril.	( )
292	Lower values of iodine number in ruminants reflect hydrogenation of ingested fat by rumen microbes.	( )
293	When two double bonds are separated by one or more CH groups, the unsaturated is said to be unconjugated.	( )

294	When both carbon chains are on the same side of the double bond of unsaturated fatty acids, the form is termed cis.	( )
295	Haemoglobin and myoglobin are conjugated proteins and latter imports oxygen in the blood.	( )
296	PSE is more common in stress susceptible animals.	( )
297	Thaw rigor much more severe than cold shortening.	( )
298	Large ice-crystal formation is generally observed in fast freezing.	( )
299	Animal lipids are mainly neutral lipids.	( )
300	PSE is normally observed in pork.	( )
301	Meat is a good source of phosphorous and iron but low in calcium.	( )
302	Calpain is responsible for fast disruption of Z-line during ageing.	( )
303	Red muscles are more susceptible to cold shortenings.	( )
304	Hanging on aitch bone is generally practiced for cold carcasses.	( )
305	High voltage electrical stimulation is normally applied after 30 min of slaughter.	( )
306	In microwave cooking 915 MHz frequency is better than 2450 MHz frequency.	( )
307	Meat colour is expressed in terms of hue, chroma and value.	( )
308	Sensory evaluation laboratory must be maintained at 30 ± 2^0C.	( )
309	Troponin is calcium ion receptive globular protein.	( )
310	Phospholipids are generally found in muscle tissues in form of phosphoglycerides.	( )
311	Generally lean meat is low in cholesterol but veal contains more cholesterol.	( )
312	It is necessary that all susceptible pigs will exhibit PSE pork.	( )
313	Pasture fed beef is good source of omega-3-fatty acid.	( )
314	Carotene is almost absent in buffalo meat.	( )
315	Chloride ion present in salt directly acts on microbial cells.	( )
316	Curing sugar may be the cause of maillard reaction.	( )
317	Liquid smoke can directly be incorporated in meat emulsion.	( )
318	During contraction of muscles, thick and thin filaments slide towards each other and decrease the sarcomere length.	( )
319	Z-line in white muscles is more labile during conditioning.	( )
320	As a general rule only lean meat is suitable for drying.	( )

Q.4 Match the following

A.

A		B		
1.	Actomyosin	a.	Tradition meat products of Kashmir	( )
2.	Cathepsin	b.	Meat sorting or grouping	( )
3.	Yakini	c.	Irradiation	( )
4.	Grading	d.	Commercial sterility	( )
5.	Cold sterilization	e.	Rigor mortis	( )
6.	Canning	f.	Conditioning	( )
7.	Psychrophites	g.	20-45^{0}C	( )
8.	Mesophils	h.	Mucor	( )
9.	Wiskers	i.	Taste and odour	( )
10.	Flavour	j.	<20^{0}C	( )

B.

A		B		
1.	Green rot	a.	Serratia sp.	( )
2.	Red rot	b.	Pre-rigor meat	( )
3.	Cold shortening	c.	High ultimate pH	( )
4.	DFD	d.	Pork	( )
5.	PSE	e.	*Pseudomonas fluorescens*	( )
6.	Jowl	f.	Sharp fall in pH	( )
7.	Round	g.	Meat spoilage	( )
8.	ERV	h.	Beef	( )
9.	Tenderness	i.	Tradition meat products of Kashmir	( )
10.	Gostaba	j.	Papain	( )

C.

A		B		
1.	Flavoprotein	a.	Desmosine	( )
2.	Potassium sorbate	b.	Carcinogen	( )
3.	Slow freezing	c.	*Serratia marcescens*	( )
4.	Bologna	d.	Pig carcass	( )
5.	Reddish discolouration of meat	e.	Sausage	( )
6.	Benzapyrene	f.	Mold inhibitor	( )

7.	Picnic	g.	Riboflavin	( )
8.	Elastin	h.	High drip loss	( )
9.	Antioxidants	i.	PSE meat	( )
10.	Decrease in WHC	j.	Ascorbates	( )

D.

A		B		
1.	Less penetration	a.	Lactic acid	( )
2.	Glycogen	b.	Y-rays	( )
3.	Increase in WHC	c.	UV rays	( )
4.	Proline	d.	Sarcoplasmic reticulum	( )
5.	Fenestrated Collar	e.	Phosphates	( )
6.	High penetration	f.	DFD meat	( )
7.	High ultimate pH	g.	Actin	( )
8.	Isodesmosine	h.	Elastin	( )
9.	Retorts	i.	Brown colour	( )
10.	Metmyoglobin	j.	Canning	( )

E.

A		B		
1.	12-D concept	a.	Green colour	( )
2.	Warmed over flavour	b.	Phospholipids	( )
3.	Freezer burn	c.	Mucor mucido	( )
4.	Whiskers	d.	*Penecillium expansum*	( )
5.	Sulfmyoglobin	e.	Pro-oxidant	( )
6.	Green patches	f.	Clostridium botulinum	( )
7.	UV light	g.	Surface dehydration	( )
8.	APEDA	h.	Sarcoplasmic reticulum pump	( )
9.	MAP	i.	1986	( )
10.	Marsh-Bendal factor	j.	Extended shelf life	( )

F.

A		B		
1.	M-line	a.	CFU	( )
2.	Z-line	b.	RN gene	( )
3.	Smooth muscles	c.	Five freedoms	( )

4.	Microbial population in the food	d.	Thick band	( )
5.	Animal welfare	e.	1973	( )
6.	PSE	f.	New Delhi	( )
7.	Acid meat	g.	Thin band	( )
8.	BIS	h.	Rynodiine receptor	( )
9.	MFPO	i.	GIT	( )
10.	Cold sterilization	j.	Irradiation	( )

G.

A		B		
1.	Linolenic acid	a.	Rancidity	( )
2.	M-line	b.	Gap filament protein	( )
3.	Lysosomes	c.	Ascorbic acid	( )
4.	Collagen	d.	Disodium inosinate	( )
5.	Desmin	e.	A-band	( )
6.	Cytochrome enzyme	f.	Essential fatty acids	( )
7.	Collagen synthesis	g.	Minor muscle pigment	( )
8.	Peroxidase	h.	Emulsifying agent	( )
9.	Myofibrillar protein	i.	Cross links	( )
10.	Flavour enhancer	j.	Cathepsin	( )

H.

A		B		
1.	IMP	a.	Sarcoplasmic protein	( )
2.	Maillard reaction	b.	Flavour	( )
3.	Thawing	c.	Meat tenderizer	( )
4.	Haemoglobin	d.	Carbohydrate and protein interaction	( )
5.	Bromolin	e.	Microwave	( )
6.	Cryogenics	f.	Unacceptable meat	( )
7.	Electrical stimulation	g.	Meat marbling	( )
8.	Freezer burn	h.	Good appearance	( )
9.	Sensory attributes	i.	Quick glycolysis	( )
10.	Myoglobin	j.	Quick freezing	( )

I.

A		B		
1.	Dr Lacker R.H.	a.	WHC and eating quality under influence of pH	( )
2.	Dr Honikel K.O.	b.	Theory of gap filaments	( )
3.	Dr Marsh B.B. and Dr Bendell, J.R.	c.	Observed effect of electrical stimulation for first time in 1749	( )
4.	Dr Benjamin Franklin	d.	Theory of muscle contraction	( )
5.	Nicolas Appert	e.	Cured meat	( )
6.	Salt	f.	Water holding capacity	( )
7.	Polyphosphates	g.	Father of canning	( )
8.	Haemochromogen	h.	Metmyoglobin	( )
9.	Brown	i.	Freezer burn	( )
10.	Dehydration	j.	Pink	( )

J.

A		B		
1.	ONCO boss or knoch Briton and Jarvin	a.	Splitting carcass into parts (buffalo)	( )
2.	Elther apparatus	b.	Freezing of meat	( )
3.	Blast turbo freezers	c.	Electrical stunning (sheep)	( )
4.	Lovibond colour meter	d.	Evaluation of meat texture	( )
5.	Dastron tester	e.	Evaluation of meat tenderness	( )
6.	W-B shear force meter	f.	Electrical stimulation high & low voltage	( )
7.	Wheal type dehider	g.	Movement of carcasses in slaughter hall	( )
8.	Overhead rail system	h.	Electrical stunning poultry	( )
9.	Electrified water bath/grid	i.	Dehiding in cattle	( )
10.	Circular saw carcass splitter	j.	Estimation of meat colour	( )

K.

A		B		
1.	Dark cutting	a.	Stress resistant pigs	( )
2.	Blood pH do not change	b.	Calcium pump	( )
3.	Husselbach's term	c.	Extra adenosine triphosphate	( )

4.	ATP splitting	d.	Onset of rigor mortis	( )
5.	Relaxing factor	e.	Pig	( )
6.	Disappearance of ATP	f.	Sarcoplasmic reticulum	( )
7.	pH 6.2	g.	Decrease in WHC	( )
8.	Perigo factor	h.	Cold sterilization	( )
9.	Lypholization	i.	Nitrite	( )
10.	Ionizing radiation	j.	Freeze drying	( )

L.

A		B		
1.	Myomesin	a.	'A'band	( )
2.	Unstable emulsion	b.	'I'band	( )
3.	Anisotropic	c.	M-line (Filamine)	( )
4.	Isotropic	d.	Non-phosphorylated lipid	( )
5.	Malonaldehyde	e.	Short meat	( )
6.	Cerebrosides	f.	TBA value	( )
7.	Desmin	g.	Cured meat colour	( )
8.	Sphingomyelin	h.	Calcium binding protein	( )
9.	Nitrosylhemochrome	i.	Phospholipid	( )
10.	Calsequestrin	j.	Z-line	( )

M.

A		B		
1.	Carbonyls	a.	Estimation of myoglobin	( )
2.	Triglycerides	b.	N-line protein	( )
3.	Cyanmetmyoglobin	c.	Esters of monohydroxy aliphatic alcohol	( )
4.	Nebulin	d.	Present both in actin and myosin	( )
5.	F-protein	e.	Acidic protein activates myosin light chain kinase	( )
6.	Waxes	f.	Binds to calcium	( )
7.	3-methyl histidine	g.	Fats (Fatty acid esters)	( )
8.	Calmodulin	h.	Binds to myosin	( )
9.	Troponin	i.	Binds very strongly with LMM	( )
10.	C-protein	j.	Meat flavour	( )

N.

A		B		
1.	Troponin binds calcium ions and forms	a.	For connective tissue	( )
2.	Troponin-T binds	b.	In actin, actinins, troponin C and H meromyosin	( )
3.	Troponin-I inhibits	c.	An equimolar complex with troponin -I	( )
4.	The higher content of aromatic residues are observed	d.	Actomyosin ATPas	( )
5.	Hydroxyprolein content of muscle is measuring stick	e.	Troponin -C	( )
6.	The concentration of Ca++ ions in the sarcoplasm is restored to rest level by reabsorption through	f.	Do not contribute to meat flavour	( )
7.	Intact fibrillar and sarcoplasmic proteins as such	g.	Sarcoplasmic reticulum pump	( )
8.	Large crystal	h.	-30^{0}F	( )
9.	Quick freezing	i.	212^{0}F	( )
10.	Acid food	j.	Slow freezing	( )

O.

1.	Intramuscular fat	a	Potato starch	( )
2.	Red spot in meat	b	Marbling	( )
3.	Potassium sorbate	c	Xanthene	( )
4.	Binder	d	Mould inhibitor	( )
5.	Stabilizer	e	*Serratia marcecens*	( )
6.	Freezer burn	f	Nylon-6	( )
7.	Infertile latebra	g	Dehydration & discoloration	( )
8.	Boil in bag packaging	h	Blastodisc	( )
9.	Thick filament	i	I-band or actin filament	( )
10.	Thin filament	j	A-band or mysosin filament	( )

Fresh Meat Technology Key

Q.No. 1					
1.b	33.a	65.a	97.b	129.d	161.c
2.d	34.a	66.b	98.a	130.c	162.a
3.a	35.a	67.b	99.d	131.c	163.b
4.d	36.c	68.b	100.c	132.c	164.c
5.a	37.a	69.a	101.a	133.d	165.a
6.c	38.d	70.a	102.b	134.a	166.a
7.b	39.c	71.b	103.a	135.b	167.a
8.a	40.b	72.b	104.d	136.d	168.c
9.a	41.d	73.a	105.a	137.c	169.a
10.d	42.c	74.d	106.d	138.d	170.c
11.a	43.a	75.c	107.a	139.a	171.b
12.d	44.a	76.b	108.b	140.a	172. c
13.a	45.c	77.c	109.a	141.c	173. a
14.b	46.c	78.b	110.c	142.a	174.a
15.b	47.a	79.a	111.c	143.d	175. b
16.a	48.b	80.d	112.d	144.d	
17.a	49.d	81.a	113.d	145.c	
18.a	50.a	82.a	114.c	146.a	
19.d	51.b	83.c	115.a	147.c	
20.a	52.c	84.a	116.c	148.c	
21.c	53.c	85.c	117.d	149.a	
22.d	54.a	86.d	118.c	150.c	
23.a	55.a	87.d	119.d	151.c	
24.a	56.c	88.d	120.d	152.a	
25.b	57.c	89.b	121.d	153.a	
26.b	58.b	90.c	122.c	154.c	
27.a	59.a	91.c	123.c	155.c	
28.a	60.a	92.a	124.c	156.c	
29.a	61.a	93.c	125.d	157.d	
30.a	62.c	94.c	126.c	158.c	
31.d	63.d	95.b	127.d	159.a	
32.b	64.b	96.d	128.c	160.a	

Q.No. 2			
1.	6.9	31.	23.5
2.	Mesodermal	32.	-0.5 to -3.0
3.	High biological value	33.	200ppm
4.	4-6	34.	Joint Secretary, Ministry of Food Processing Industry
5.	Kidney close side	35.	Brown
6.	DNA Fragmentation	36.	H-band
7.	Log	37.	75-80%
8.	E.coli	38.	Trypsin
9.	White	39.	Cofilin
10.	Sarcomere	40.	10-100 mμ
11.	Vitamin-C	41.	Four
12.	Actomyosin bond	42.	Two
13.	Marbling	43.	Collagen
14.	85-90%	44.	Fenestrated Collar
15.	Lag phase	45.	Higher
16.	>0.99	46.	Red
17.	Meat Food Products Order	47.	Formaldehyde
18.	Hydroxyproline	48.	Actin filament in each half sarcomere
19.	Unsaturated	49.	Cured
20.	Mutton	50.	Nitric oxide
21.	Highest	51.	Skeletal
22.	Monosodium Glutamate	52.	Muscle cells
23.	Myosin	53.	Food Safety and Standards Authority of India
24.	Lactic acid	54.	Sarcomere
25.	Category-A	55.	Actomyosin bond
26.	Oidium carnis or Cladosporium herbarum	56.	Speed up of rigor mortis
27.	Sulfur	57.	<5%
28.	Metal foil laminates	58.	Structural or stromal
29.	Phenol	59.	Freezing
30.	Eight	60.	Nitrosamine

61.	5-7^{0}C	91.	Chemical resistance properties
62.	Thermoplastic films	92.	Non fermentative G-ve bacteria
63.	72	93.	More Collagen
64.	Potato Dextrose Agar	94.	More
65.	Pseudomonas mephatica	95.	Less developed respiratory mechanism
66.	50 to 70	96.	Water soluble proteins
67.	376	97.	35%
68.	60	98.	Decrease
69.	6	99.	Less fat content
70.	2006	100.	Less
71.	Sarcomere	101.	Staphylococcus aureus
72.	Chatol	102.	Glycolytic
73.	Andrew Boss	103.	80%
74.	Kidneys	104.	Z
75.	Diphysis	105.	C-protein
76.	35-55	106.	20-50
77.	1-10°C	107.	Mg^{2+} ATP
78.	0.5-1.0	108.	Tensile
79.	2.0	109.	Pigment
80.	77^{0}C	110.	More
81.	Age	111.	Oxygen
79.	2.0	112.	Heating
80.	77^{0}C	113.	-20^{0}C
81.	Age	114.	Feathering
82.	Time in minutes	115.	Ground substances
83.	Thermal death time	116.	10^{-7} moles/ltr
84.	0.3-2.0	117.	10 to 100
85.	32^{0}F or 0^{0}C	118.	Homeostasis
86.	Cellulose derivatives	119.	Increase
87.	Epimysium	120.	60%
88.	15^{0}C	121.	Cold shortening
89.	-1 to -4	122.	Stearic effect
90.	Bitter	123.	Lysozyme

124.	High	158.	Ferric
125.	5^{0}C	159.	0.30hrs
126.	45-60^{0}C	160.	35-65%
127.	3Kgy	161.	Cardiac
128.	15-20^{0}C	162.	Involuntary
129.	PSE condition	163.	600
130.	High	164.	Muscle fibre
131.	Biphasic	165.	Sarcolemma
132.	Accelerating	166.	Regulatory muscle proteins
133.	2	167.	50-55%
134.	Refrigerator	168.	Intermuscular or seem
135.	-18^{0}C	169.	Dressing percentage
136.	N_2 and CO_2	170.	pH
137.	-19^{0}C	171.	Rise
138.	0.99	172.	Delay
139.	0.995 to 0.990	173.	60%
140.	0.94	174.	Cold shortening
141.	0.86	175.	80-90%
142.	TDT	176.	Water holding capacity
143.	Pigs	177.	Cross bridge
144.	Cattle	178.	Psoas
145.	Instron	179.	Loin and round
146.	92%	180.	Triceps brachii
147.	2	181.	85-90%
148.	Increased risk of inbreeding	182.	Tallowy or chalky
149.	Mea production	183.	Gradually
150.	Beef	184.	Minimum
151.	Pork	185.	Minimum
152.	300	186.	100%
153.	Ammonical silver solution	187.	Onset
154.	Intermolecular cross links	188.	1/4
155.	Less	189.	3
156.	Loss of WHC	190.	6
157.	Contractile actomyosin ATPase	191.	Higher

192.	Oxymyoglobin	228.	Gel Point
193.	Isoelectric point	229.	10ppm
194.	Oxymyoglobin	230.	20-100
195.	Nitric oxide haemochromogen	231.	Carbonyls
196.	5-alpha androstenon-16ene –zene	232.	Syneresis or weeping
197.	75%	233.	02
198.	1840	234.	At an angle to the longitudinal axis of the muscle.
199.	Nicolas Appert	235.	Field of Cohenheim
200.	Drip	236.	Myofibrillar proteins
201.	Freezer burn	237.	Ribosomes aggregates
202.	1813	238.	4,60,000
203.	1955	239.	1.6 µm
204.	Chlortetracyclene	240.	24mg/g of muscle
205.	Sodium benzoate	241.	Oxidative metabolism
206.	Yeast	242.	400
207.	Fungistatic	243.	Oleic acid
208.	7.0	244.	Coalescence
209.	Shrinkage	245.	5-alpha androst
210.	Roentgen	246.	Methyl mercapton
211.	Radicidation	247.	Endomysium
212.	2-6	248.	Connectin (Ertitin)
213.	Quick or fast	249.	Actin
214.	Increasing rate	250.	Free Redical
215.	Z-value	251.	Adenylate deaminase
216.	10 kGy	252.	Black
217.	Thermophiles	253.	Brown
218.	15-20	254.	Criss cross
219.	Resistance to ionizing radiation	255.	Non
220.	Allicin	256.	Controitin sulphate
221.	0.6-0.84	257.	Sarcomere
222.	Natamycin	258.	Each
223.	5%	259.	Three
224.	Denaturation	260.	Z-line
225.	Desmin_ and _connectin	261.	T-system
226.	13%	262.	Salt and ATP
227.	0.3%	263.	Chromophoric

264.	1.6	278.	Cladosporium
265.	Bright	279.	Sporotrichum carnis
266.	Methane carbon	280.	Suseptible
267.	Oxidation	281.	RN or Napole gene
268.	Fat	282.	Metmyoglobin
269.	Tropocollagen	283.	85-90%
270.	Phosphocreatinine	284.	Cherry red colour
271.	45-65	285.	Tropomyosin-B
272.	Metmyoglobin	286.	Calcium
273.	85-90%	287.	Pig
274.	Chondroitin So_4	288.	Z-line
275.	Higher	289.	2005
276.	10^7 bacteria per gram	290.	2-3 rib
277.	One		

Q. No. 3							
1.T	41.F	81.F	121.T	161.T	201.F	241.F	281.T
2.T	42.T	82.F	122.T	162.F	202.T	242.F	282.T
3.F	43.T	83.T	123.T	163.T	203.T	243.F	283.T
4.T	44.T	84.F	124.T	164.F	204.F	244.T	284.T
5.F	45.T	85.T	125.F	165.F	205.F	245.T	285.T
6.T	46.F	86.F	126.F	166.F	206.T	246.T	286.T
7.F	47.T	87.T	127.T	167.T	207.F	247.T	287.T
8.T	48.F	88.F	128.T	168.T	208.T	248.F	288.T
9.F	49.F	89.F	129.T	169.T	209.F	249.T	289.T
10.F	50.T	90.T	130.T	170.F	210.F	250.F	290.T
11.T	51.T	91.T	131.T	171.F	211.T	251.F	291.T
12.T	52.F	92.T	132.F	172.T	212.T	252.F	292.T
13.T	53.F	93.F	133.F	173.F	213.T	253.T	293.T
14.F	54.T	94.F	134.F	174.T	214.F	254.F	294.T
15.F	55.T	95.F	135.F	175.T	215.F	255.T	295.T
16.T	56.F	96.T	136.F	176.F	216.T	256.T	296.T
17.F	57.F	97.T	137.T	177.T	217.F	257.T	297.T
18.F	58.F	98.T	138.T	178.F	218.F	258.F	298.F
19.F	59.T	99.F	139.T	179.F	219.F	259.T	299.T
20.T	60.T	100.T	140.F	180.T	220.T	260.T	300.T
21.T	61.F	101.F	141.T	181.F	221.F	261.T	301.T
22.T	62.T	102.F	142.T	182.T	222.F	262.F	302.T
23.T	63.T	103.F	143.T	183.F	223.T	263.T	303.T
24.F	64.F	104.T	144.T	184.F	224.T	264.T	304.F
25.F	65.T	105.F	145.F	185.F	225.F	265.T	305.T
26.T	66.F	106.F	146.F	186.T	226.T	266.T	306.T
27.F	67.T	107.T	147.F	187.F	227.T	267.T	307.T
28.T	68.F	108.T	148.T	188.F	228.T	268.T	308.F
29.F	69.T	109.F	149.T	189.F	229.T	269.F	309.T
30.F	70.T	110.T	150.F	190.F	230.T	270.F	310.T
31.F	71.F	111.T	151.F	191.T	231.T	271.F	311.T
32.T	72.T	112.T	152.T	192.T	232.F	272.F	312.F
33.F	73.T	113.F	153.F	193.F	233.T	273.T	313.T
34.F	74.F	114.F	154.T	194.T	234.T	274.F	314.T
35.F	75.F	115.T	155.F	195.T	235.T	275.F	315.T
36.F	76.T	116.T	156.T	196.F	236.F	276.T	316.T
37.F	77.F	117.F	157.T	197.T	237.T	277.T	317.T
38.T	78.F	118.F	158.T	198.T	238.F	278.T	318.T
39.F	79.F	119.T	159.T	199.T	239.T	279.T	319.T
40.T	80.T	120.F	160.F	200.T	240.T	280.T	320.T

Q.No. 4															
A		B		C		D		E		F		G		H	
1.	e	1.	e	1.	g	1.	c	1.	f	1.	d	1.	f	1.	b
2.	f	2.	a	2.	f	2.	a	2.	b	2.	g	2.	e	2.	d
3.	a	3.	b	3.	h	3.	e	3.	g	3.	i	3.	j	3.	e
4.	b	4.	c	4.	e	4.	g	4.	c	4.	a	4.	i	4.	a
5.	c	5.	f	5.	c	5.	d	5.	a	5.	c	5.	b	5.	c
6.	d	6.	d	6.	b	6.	b	6.	d	6.	h	6.	g	6.	j
7.	j	7.	h	7.	d	7.	f	7.	e	7.	b	7.	c	7.	i
8.	g	8.	g	8.	a	8.	h	8.	i	8.	f	8.	a	8.	f
9.	h	9.	j	9.	j	9.	j	9.	j	9.	e	9.	h	9.	g
10.	i	10.	i	10.	h	10.	i	10.	h	10.	j	10.	d	10.	h

Q.No. 4													
I		J		K		L		M		N		O	
1.	b	1.	f	1.	e	1	c	1.	j	1.	c	1.	b
2.	a	2.	c	2.	a	2.	e	2.	g	2.	e	2.	e
3.	d	3.	b	3.	b	3.	a	3.	a	3.	d	3.	d
4.	c	4.	j	4.	c	4.	b	4.	b	4.	b	4.	a
5.	g	5.	d	5.	f	5.	f	5.	h	5.	a	5.	c
6.	e	6.	e	6.	g	6.	d	6.	c	6.	g	6.	g
7.	f	7.	i	7.	d	7.	j	7.	d	7.	f	7.	h
8.	j	8.	g	8.	i	8.	i	8.	e	8.	j	8.	f
9.	h	9.	h	9.	j	9.	g	9.	f	9.	h	9.	j
10.	i	10.	a	10.	h	10.	h	10.	i	10.	i	10.	i

4

Processed Meat Technology

Q.1 Multiple type questions

1. Cured meat colour is due to

   a. Reduced myoglobin
   b. Oxymyoglobin
   c. Oxidized myoglobin
   d. Nitric oxide myoglobin

2. Salt addition

   a. Increases water holding capacity
   b. Retards fat oxidation
   c. Imparts desirable flavour
   d. Improves emulsifying capacity

3. Sugar in curing mixture is supposed to

   a. Increase flavour
   b. Improve colour
   c. Improve texture
   d. Retards rancidity

4. Phosphates are known to

   a. Improve colour
   b. Enhances flavour
   c. Increases tenderness
   d. Increased water holding capacity

5. Nitrates in curing solution

   a. Retards microbial growth
   b. Improves colour
   c. Increases water holding capacity
   d. Increases emulsifying capacity

6. The principle involved in the production of intermediate moisture meat is related to

   a. Osmotic pressure
   b. Moisture content of the product
   c. Dry matter content
   d. Ratio of protein to water

7. Shorting out in cooked sausage is due to

   a. Over chopping
   b. Over heating
   c. Too much collagen
   d. All of the above

8. DFD meat as compared to PSE meat
   a. Has less reflectance
   b. Retards salt penetration during curing
   c. Both of the above
   d. None of the above
9. Smoking of meat is found to
   a. Enhance the flavour
   b. Enhance the colour
   c. Retards rancidity
   d. All of the above
10. The intermediate moisture meats are processed to
   a. a_w of 0.90 or more
   b. a_w of 0.75 or less
   c. a_w of 0.85 or less
   d. a_w of 0.85 or more
11. While smoking a meat product the amount and rate of smoke is influenced by
   a. Smoke density
   b. Surface of product
   c. Relative humidity
   d. All of the above
12. Phenols from the wood smoke have
   a. Coagulation effect on surface of meat
   b. Antioxidant effect of smoked meat
   c. No effect whatsoever on meat products
   d. All of the above
13. Monosodium glutamate is used as a curing ingredient in meat products to
   a. Stabilize the colour
   b. Improve the brine retention
   c. Prevent the growth of microbes
   d. Enhance the flavour
14. Luncheon meat is a
   a. Ready to eat finely ground poultry meat
   b. Ready to eat coarsely ground rabbit meat
   c. Ready to eat finely ground pork meat
   d. Finally minced cooked pork mixed with cereal
15. The smoking of meat
   a. Increases bacterial population
   b. Decreases bacterial population
   c. Has no effect on bacterial population
   d. None of the above

16. The best quality smoke is produced at a combustion temperature of

    a. 140-200°C b. 340-400°C

    c. 240-300°C d. None of the above

17. Intermediate moisture meats contain moisture content of

    a. 10-20% b. 30-40%

    c. 20-30% d. 40-50%

18. Curing if meat is generally done at a temperature of

    a. 38-40°F b. 48-50°F

    c. 58-60°F d. 28-30°F

19. In canning of meat, the recommended head space is

    a. 1.6mm b. 1.4mm

    c. 1.6cm d. None of the above

20. Fresh sausages have a moisture content of

    a. 40-45% b. 50-55%

    c. 65-70% d. None of the above

21. The best temperature of sausage cooking is

    a. 168°C b. 168°F

    c. 148°F d. 200°F

22. Smoking of meat alter its nutritive value

    a. Adversely b. No effect

    c. Positively d. None of the above

23. Meat analogues are generally prepared from soybean meal containing

    a. 40% fat b. 30% fat

    c. 20%fat d. 2.4% fat

24. The book "Processed Meats" published by AVI publishing Company has been written by

    a. Desrocier, N.W. b. Frazier

    c. Ledward, D.A. d. None of the above

25. Irradiation preservation of foods is being used on
    a. Pilot plant scale
    b. Experimental scale
    c. Commercial scale
    d. None of the above
26. The stability of nitro-pigment in cured meat depends on
    a. Concentration of salts in curing
    b. Concentration of nitrites
    c. Partial pressure of oxygen in the atmosphere to which pigment is exposed
    d. None of the above
27. The development of slimy surface on the cured meat is due to
    a. Heavy growth of mold
    b. Heavy growth of bacteria
    c. Heavy growth of yeast
    d. Heavy growth of bacteria and yeast
28. Cured meat products contain
    a. A few putrefactive aerobic spore forming bacteria
    b. A few putrefactive anaerobic spore forming bacteria
    c. Both of the above
    d. None of the above
29. Dehydrated fruits and vegetables are especially subject to
    a. Enzyme action
    b. Microbial growth
    c. Insects attack
    d. None of the above
30. The freeze dried products are
    a. Superior to air dried products
    b. Superior to fresh foods
    c. As good as air dried products
    d. None of the above
31. Dehydrated products are normally packed in
    a. Rigid containers
    b. Flexible pouches
    c. Lined cartons
    d. All of the above
32. I bands are also known as
    a. Intervening disc
    b. J-disc
    c. Q-bands
    d. None of the above

33. One sarcomere length in a myofibril is the distance between

a. Two M-lines
b. Two A-bands
c. Two Z-lines
d. None of the above

34. The central T tubules and two tubular elements of the terminal cisternae collectively form a structure known as

a. Triad
b. Fenestrated collar
c. L-band
d. None of the above

35. Temperature of comminuted meat products at the time of sensory evaluation is

a. 60-70 °C
b. 70-80 °C
c. 40-60 °C
d. 30-35 °C

36. Write the example of dried sausage

a. Thuringer
b. Frankfurters
c. Bologna
d. Salami

37. Optimum age of chicken for preparation of tandoori is

a. Eight weeks
b. Six weeks
c. Four weeks
d. Ten weeks

38. Gradual decrease in odour or flavor is observed during frozen storage due to

a. Oxidative rancidity
b. Shrinkage
c. Bloomimg
d. Microbial activity

39. Canning was developed by

a. A. Flemings
b. A. Springer
c. Nicolas Appert
d. None of the above

40. Preservative effect of dehydration is due to reduction of____________ in meat.

a. Water activity
b. Moisture
c. Both of the above
d. None of the above

41. Level of phosphate should not exceed________ in finished meat product.

a. 0.5%
b. 1.0%
c. 1.5%
d. 2.0%

42. Ash content of cured meat is _________ than fresh meat

a. Higher
b. Lower
c. Almost equal
d. None of the above

43. Mold growth in dehydrated meat occurs when moisture content exceeds

a. 5%
b. 10%
c. 15%
d. 20%

44. Radiation dose of 4-10 megarad produces__________________smell in meat

a. Wet dog
b. Hot dog
c. Piggy smell
d. Catty smell

45. Process by which bacterial growth is prevented by simultaneous use of many factors like pH, temperature, water activity is referred as

a. HACCP
b. Hurdle Technology
c. Flash Technology
d. None of the above

46. Edible coating of meat and meat products in common term is referred as

a. Inner coating
b. Broiling
c. Enrobing
d. None of the above

47. Polyethylene can be safely used upto a temperature of

a. 70°C
b. 50°C
c. 80°C
d. 100°C

48. Polyamide film is called______________in packaging trade.

a. Nylon
b. Pterepthalate
c. Teracota
d. None of the above

49. Parchment paper wrapping is especially suitable for meat cuts with more

a. Protein
b. Fat
c. Vitamin
d. Minerals

50. High pressure technology of meat processing generally using ___________ pressure

a. 800-1200MPa
b. 100-600 MPa
c. 20-50 MPa
d. None of the above

51. Indicate the examples of dry sausages

a. Salami
b. Pepperoni
c. Both of the above
d. None of the above

52. Indicate the examples of semi-dry sausages

a. Bologna
b. Fermented meat products
c. Both of the above
d. None of the above

53. ___________rays are commonly used for preservation of foods including meat and meat products

a. Gamma rays
b. Beta rays
c. Alpha rays
d. None of the above

54. Radiation from _______is cheaper and has greater penetrating power than electrons

a. Cobalt-60
b. Ceseum-137
c. Pluton-213
d. None of the above

55. Refrigeration life of redurised fresh meat is

a. 3-4 days
b. 5 weeks
c. 8 weeks
d. 10 weeks

56. Radicidation is capable to destroy parasitic reproduction at the dose of

a. 1-2 KGy
b. 2-4 KGy
c. 0.1-1 KGy
d. 5-8 KGy

57. Radappertized or cold sterilized meat is generally stored

a. Refrigeration
b. Freezing
c. Shelf stable
d. None of the above

58. Indicate the role of sorbates in meat preservation

a. Mold inhibitors
b. Yeast inhibitors
c. Both of the above
d. None of the above

59. Indicate the role of organic acids in meat preservation

a. Mold inhibitors
b. Yeast inhibitors
c. Acidulant
d. None of the above

60. Hurdle technology

    a. Heat
    b. Refrigeration
    c. Irradiation
    d. Combinations of all these

61. Indicate the example of fermented sausage

    a. Thuringer
    b. Cerevalat
    c. Both of the above
    d. None of the above

62. Lacter bacteria produces____________flavour in fermented sausages

    a. Bitter
    b. Sweet
    c. Tangy
    d. None of the above

63. Indicate the function of addition of cold water or ice during chopping

    a. Proper texture
    b. Reduction of heat during chopping
    c. Better emulsion
    d. All of the above

64. End product of non-enzymatic browning in processed food is

    a. Deladoinine
    b. Furfural
    c. Both of the above
    d. None of the above

65. Liquid smoke consists of

    a. 25parts liquid smoke
    b. 5 parts citric acid
    c. 65-75 parts water
    d. All of the above

66. WOF is never observed in

    a. Fresh meat
    b. Cooked meat
    c. Cured meat
    d. None of the above

67. Name commonly used colour accelerators which acts as reducing agents are

    a. Ascorbates
    b. Erythorbates
    c. Both of the above
    d. None of the above

68. Indicate the example of semidry sausage

    a. Thuringer
    b. Cerevalat
    c. Both of the above
    d. None of the above

69. Severe heating such as grilling produces traces of carcinogens such as

    a. Benzpyrene
    b. Dibenzanthracins
    c. Both of the above
    d. None of the above

70. Name the compound exhibits both antioxidant and bacteriostatic effect in smoking of meat

    a. Phenol
    b. Formaldehyde
    c. Carbonyls
    d. None of the above

71. Ascorbic acid and Polyphosphate acts as

    a. Colour developers
    b. Flavour enhancers
    c. Curing adjuncts
    d. None of the above

72. Prolonged storage of irradiated raw meat at temperature 60 - 100°F is responsible for

    a. Bitter flavour
    b. Tyrosine crystals isolation
    c. Both of the above
    d. None of the above

73. The two commonly used frequencies (910 and 2450) of heating are used in

    a. Microwave heating
    b. Pulse electric field
    c. Electric kettle heating
    d. None of the above

74. Stability of meat emulsion depends upon

    a. Chopping
    b. Short meats
    c. Salt
    d. All of the above

75. Severe heating such as grilling produces traces of carcinogens such as

    a. Benzpyrene
    b. Dibenzanthracins
    c. Both of the above
    d. None of the above

76. Canadian bacon is prepared from

    a. Pork loin muscle
    b. Strip muscle
    c. Sirloin muscle
    d. All of the above

77. Triple point of water exists at

    a. 4.7 mmHg pressure & 25°F temperature
    b. 5.0 mmHg pressure & 20°F temperature
    c. 4.0 mmHg pressure & 15°F temperature
    d. All of the above

78. Freezing and salting are ways of preserving desirable characteristics in ____ during storage

a. Cold boned meat
b. Cured meat
c. Hot boned meat
d. All of the above

79. Most meat curing is confined to

a. Beef
b. Pork
c. Beef & Pork both
d. None of the above

80. Name the colour fixation agents in meat curing

a. $NaNO_3$
b. KNO_3
c. Both of the above
d. None of the above

81. Match the correct pair for tongue taste buds

a. Tip of tongue for sweet/salty taste
b. Middle of tongue for sour taste
c. Base of tongue for bitter taste
d. All of the above

82. Green rooms are commonly maintained at

a. 79-82°F and 80-90% RH
b. 80-82°F and 70-80% RH
c. 70-80°F and 85-95% RH
d. None of the above

83. The starter culture for dry sausages consists of

a. Lactobacillus
b. Leuconostoc
c. Micococcus
d. All of the above

84. A non-enzymatic browning reactions between sugar and amino acids is essential for development of characteristics

a. Meat Colour
b. Meat flavour
c. Meat binding
d. None of the above

85. Sterility of canned meat can be checked by incubating at

a. 30°C and 50°C
b. 37°C and 55°C
c. 57°C and 85°C
d. None of the above

Q.2 Fill in the blanks

1. _______________is a typical example of fine-cut sausage extruded into large casings.
2. The temperature of meat when taken from the cutter should be about _______________.
3. _______________ is an example of cooked smoked sausage.
4. Excessive shrinkage during smoking may produce _______________ appearance.
5. Hot water at approximately _______________ is sprayed for cooking franks.
6. Fermented sausages such as thuringer, cerevalat make use of lacter bacteria which produce ____________ flavour.
7. Myoglobin has maximum absorption at wavelength of _______________.
8. NID regulations states that cooked sausages may not contain more than _______________ added water or moisture
9. Hams for canning shall not be more than_____________ water over and above cured ham.
10. One way to avoid green rings and cores is to ensure that the internal temperature of the product reaches_____________________.
11. ____________________is responsible for desired colour of fresh meats.
12. Use of phosphates is restricted to not more than _______________ in the finished product.
13. Development of meat flavour in heated meat is due to interaction between hypoxanthene and ___________________________.
14. The process _________________for production of meat analogues is carried out at isoelectric pH.
15. In finely comminuted meat the continuous structure is made up of ___________________.
16. ____________________________is the desired pigment in heat processed cured meats.
17. Sodium chloride when 2 to 8% in aqueous phase of the meat product acts as a _________________ conditioner.
18. While mincing meat for frankfurter type sausages the destruction of ________________ leads to increased WHC.

19. In sausage emulsion, meat proteins act as__________________.
20. Fat caps appear in sausages due to _______________ (protein) breakdown.
21. Intermediate moisture meats prepared by________________technique are equilibrated in humectants solution of low a_w.
22. Non enzymatic browning of intermediate moisture meats resulting into formation of an in soluble pigment called _______________________.
23. In canned meats during retorting operation_______________is released resulting into black stained tinplates.
24. Fibers produced by extrusion through an orifice are _________________ protein fibers.
25. The taste panel required for new product development is called _________.
26. During frying the formation of __________________ from nitrosamines in processed meat is a major problem.
27. To set up a moisture constraint for ingredients with different protein to moisture ratios the equation is_________________________________.
28. Tangy flavour in fermented sausages is due to__________________________.
29. Amount of myoglobin is _______________ proportional to muscular activity of the tissue.
30. Nitrate serves principally as a source of_____________________.
31. Cooked sausage colour is developed through a process known as _____________.
32. _____________viscosity emulsions are more subject to physical disruption during processing.
33. Cutting of meat into small particles is called________________________.
34. ____________________film is used for producing cook-in hams and poultry products.
35. All pasteurized canned products must be labeled as ________________.
36. Alarm water content in fat free dehydrated meats is ________________.
37. _____________________is an example of dry sausage.
38. Electric eye instrument is used to measure the smoke density which measures the smoke density in terms of___________________________.
39. The pH of semidry sausages is in between ____________________.

40. Lactobacillus and Leuconostoc cultures are used for making fermented sausages and average processing and ripening period is approx. __________.

41. Restructured meats are more susceptible to__________________ during storage.

42. Average pH of hot boned beef is____________________.

43. Intermediate moisture pet food is made by __________________ technique.

44. ____________________is an expression of the desirable water present in a food products.

45. Can used for meat should be lacquered with ________________ resistant compounds.

46. Warmed over flavour develops in pre-cooked refrigerated meat products is due to__________________.

47. A binder is rich in ____________________.

48. Filler consists mainly of ____________________.

49. Frankfurters are ____________________ type emulsion.

50. Water holding capacity of pre-rigor meat is__________________.

51. Sheep casings are an example of ____________________casings.

52. Denaturation of proteins _________________the water holding capacity of meat.

53. Salt should be added in the ____________________ of sausage making.

54. Massaging and tumbling helps in extraction of proteins which are useful in_______________meat chunks.

55. _________________are salt soluble proteins.

56. Fresh sausages are___________________ type.

57. The most common Indian meat preparation is__________________.

58. The pink colour of cured meat products develops on__________________.

59. The aroma of a product can be perceived by ____________________.

60. Desirable glosses produced in smoked meat due to production of ________ ____________________.

61. Emulsion temperature should not exceed _____________ in the manufacture of sausage.

62. Cans used for meat should be __________________ lacquered.

63. ________________ is an example of putrefactive anaerobe important in thermal process evaluation for meats.

64. Salt contents of fresh and fermented sausages____________ respectively.

65. Nitrite prevents the growth of ____________________organism.

66. Nitrosamines are produced by reaction between ____________ and secondary amines.

67. MSG and hydrolised vegetable proteins enhance____________ of cured meat products.

68. Maximum residual level of nitrite allowed in various cured meat products is_______ppm.

69. ________________take part in reduction of metmyoglobin to myoglobin.

70. Myoglobin has a _______________ affinity for oxygen than haemoglobin.

71. Pickle is injected into _________________artery of ham to have uniform distribution.

72. ________________brine is most common under commercial conditions.

73. Pumping schedule commonly calls for adding ____________by weight of pump pickle.

74. Formaldehyde has only______________ act.

75. Temperature of 750^{0}F in smoking favours the formation of ____________ which is carcinogenic.

76. Liquid smoke is generally prepared by_____________of hardwood sawdust.

77. Cooking______________red colour in cured meats.

78. Temperature and humidity during curing should be maintained at__________ and _______________respectively.

79. Massaging and tumbling helps in extraction of ____________which helps in binding of emulsion.

80. Tough meat gives best results if cooked by_________________________.

81. ________________ratios of various tissues are important in preparing sausages formula.

82. More than ____________level of mechanically deboned meat in sausages have adverse effects.

83. Sectioning and forming allows ____________ compositional control.

84. Semidry sausages have a yield of about_____________ of original weight.
85. Dry sausages have a yield of _______________ of original weight.
86. A pH of 5.3 or less provides _______________flavour to semidry sausages.
87. Corning refers to preservation of beef by___________________.
88. Storage place for cans may not have more than __________________RH.
89. Pasteurized meats are cooked to an internal temperature of______________.
90. Surface greening is not noticeable for at least ________________ days after processing.
91. Water that is unavailable for processing is called_____________________.
92. Greening in sausage is caused by _________________________ bacteria
93. Greening in canned ham is caused by_____________________________.
94. Packaging materials for cured meats should have __________________ oxygen permeability.
95. _________________________ are curing adjuncts.
96. Low acid canned foods have pH above__________________.
97. __________________ is the only ingredient necessary for curing of meat.
98. PSE in meat science stands for ________________________________.
99. In the emulsion type sausages ________________ and stabilized by the lean components.
100. Semidry sausages are characterized by a moisture content of __________.
101. Permitted moisture content in cooked sausage is__________________.
102. The rate of sausages smoking is influenced to a large extent by the _______________ of smoke house.
103. Permitted moisture content in raw sausages is_______________________.
104. _________________is added as substrate to meat mixture for acid production.
105. Ability of meat proteins to bind water is _______________ at isoelectric point.
106. Semidry sausages have a finished moisture content of_________________.
107. Semidry and dry sausages are course ground and _________________ type products.

108. Dry sausages have a moisture content of about ____________________.

109. ____________________ is an example of dry sausage.

110. ____________________ and semidry sausage.

111. *Candida utilis, Saccharomyces cerevisiae, Saccharomyces uvarum* and *Kluyveromyces fragilis* are the inactive dry yeasts often included in fabricated foods at level of ____________________.

112. ____________________ is the bacteria used for commercial production of biomass.

113. Protein content is highest in ____________________ biomass.

114. Protein content is lowest in ____________________ biomass.

115. Fish sauce is known as ____________________ in Phillipines.

116. Fish paste are known as ____________________ in Phillipines.

117. Fermentation time for fish sauce varies with ____________________ of fish.

118. In general more the concentration of salt, more the ____________________ are hydrolysed.

119. The aroma of fish sauce is a blend of ____________________.

120. Organism used in bulgarian milk production is ____________________.

121. Neutralization reduces loss of ____________________.

122. Neutralization control ____________________ flavour.

123. Neutralization increases ____________________ quality.

124. Meat curing was developed to reduce the need for ____________________.

125. ____________________ is a combined response of aroma (smell) and taste.

126. Cans used for meat should be lacquered with ____________________ resistant compounds.

127. Meat is a ____________________ acid food.

128. ____________________ is an example of putrefactive anaerobe important in thermal process evaluation.

129. Persons suffereing from cold ____________________ participate in taste panel.

130. Cured meat can be an important source of ____________________ that are carcinogenic.

131. Microbiological standards for precooked sausages is ____________________ of viable aerobes, absence of coliforms from a sample of 0.1g and total absence of pathogenic organisms.

132. The maximum limit of coal tar colours which may be added to food should not exceed ________________/kg of finished food.

133. Declaration of________________ of manufacture is necessary in prepackaged meat containers as per prevention of food adulteration act,1955.

134. Sausages containing raw meat, cereals and condiments should not contain sulphur dioxide as a preservative beyond ______________, as per prevention of food adulteration act,1955.

135. As per PFA act (1955), gelatin should not have sulphur dioxide more than _____________.

136. In ______________ concept a critical limit is a value which separates acceptability from unacceptability.

137. Commercial sterility is defined as the condition in which all ____________________ bacterial spores alongwith other pathogenic bacteria are destroyed.

138. Commercial sterility can be achieved at __________________.

139. AMSA referred sample size for standard sensory evaluation of meat products is ________________.

140. Flavour can be sensed by the human nose at ________________________.

141. Dry sausages contain water____________________________.

142. Semi-dry sausages contain water_______________________.

143. Water activity of fresh meat is____________________________.

144. Water activity of fermented sausages is_______________________.

145. Water activity of Intermediate Moisture Meat products is ______________.

Q.3. Write true or false against each statement

1.	In a roasted chicken the development of brown colour is due to enzymatic browning.	( )
2	Collagen starts softening at 70^{0}C.	( )
3	Cured meat colour stability is affected by light.	( )
4	Conversion of myoglobin to met-myoglobin is maximum at oxygen tension of 4 mmHg.	( )
5	H_2S is produced in heated meat due to breakdown of sulphur containing amino acids.	( )
6	Salt additions reduce cook loss in hams.	( )
7	Sugar helps growth of desirable bacteria during curing.	( )
8	Sodium erythorbate is found to enhance colour production and stability.	( )

9	Antioxidant effect of smoking is due to phenols present in smoke.	( )
10	PSE meat has low reflectance.	( )
11	WHC is minimum at isoelectric pH.	( )
12	RH in smoke house should be about 60%.	( )
13	NO reacts more rapidly with reduced myoglobin than metmyoglobin.	( )
14	Too much collagen leads to stickiness in sausages.	( )
15	Myosin is mostly responsible for emulsifying capacity of meat.	( )
16	Increased toughness in heated meat is due to coagulation of muscle fibre protein.	( )
17	Hot meat has better emulsifying properties and decreased cooking loss.	( )
18	In pre-rigor salted meat the saltiness is less due to poor salt absorption.	( )
19	Presence of oxygen leads to better colour stability of cured meats.	( )
20	Sausages are oil -in –water emulsions in which fat forms the discontinuous phase.	( )
21	Fillers have higher protein contents than binder.	( )
22	Intermediate moisture meats prepared by desorption contain more water than those prepared by adsorption.	( )
23	It is desirable to extract salt soluble proteins while comminuting meat for dry and semidry sausages.	( )
24	Ascorbic acid is a curing adjunct which helps in improving colour of cured meat pigments.	( )
25	Nitrosyl hemochrome and denatured metmyoglobin can be oxidized to porphyrin.	( )
26	Summer sausages fall in the group of dry sausages.	( )
27	Protein filaments are formed by passing dispersed proteins through spinnerst into a coagulating bath of an acid salt solution.	( )
28	For a new meat product development expert profile panels are analytical tools for sensory evaluation.	( )
29	The central atom of myoglobin contributes six pairs of electron and accepts nil.	( )
30	Mixers are suitable for large pieces of meat under vacuum.	( )
31	Tumbling helps the protein matrix to accumulate on the surface of meat.	( )
32	Multiformula model is used to determine least cost formula of one product.	( )

33	Least cost causes formulation never has any constraints.	( )
34	Smoke alters generation partitions into vapour and particle state.	( )
35	Smoking results into destruction of niacin and riboflavin.	( )
36	Colour is lost by dry salt curing.	( )
37	Water soluble proteins have no role in sausage emulsification.	( )
38	Old and culled animals are not suitable for comminuted meat products.	( )
39	Pig meat is a low acid food.	( )
40	Cured meats are highly susceptible to oxidative rancidity.	( )
41	Cured meats can be important source of nitrosamines that are carcinogenic.	( )
42	Greenish discolouration in sausages is caused by PA3679.	( )
43	Fermented meat products become rich source of vitamins-C.	( )
44	Comminuted meat products need very little cooking as compared to intact muscles.	( )
45	Addition of polyphosphates in meat products lowers the cooking losses.	( )
46	*Clostridium botulinum* is a mesophilic bacteria.	( )
47	About 20% of the meat produced in India is sold as processed meat.	( )
48	Filler and binders are added towards the end in sausage preparation.	( )
49	Desirable gloss is produced in smoked meats due to sodium chloride.	( )
50	Flaked and formed meat products are emulsion type.	( )
51	Only low moisture content is responsible for long shelf life of meat pickles.	( )
52	There is a danger of case hardening in dehydration of meat.	( )
53	Keeping out oxygen helps prevent oxidative rancidity.	( )
54	Galvenised iron vessels can be used to handle meat products.	( )
55	Maximum water content of sausages is not controlled by law.	( )
56	The structure forming properties of non-meat proteins are important.	( )
57	Mechanically deboned meat is suitable for emulsified products such as frankfurters, meat loaf and bologna.	( )
58	Liver and other organ meats are generally lower in B-complex content than muscle.	( )
59	Meat is a good source of calcium and low in phosphorus.	( )

60	Phosphates are added to cure to decrease the water retention capacity of meat.	( )
61	Liquid smoke is generally produced from soft wood.	( )
62	Denaturation and coagulation of proteins during cooking of meat increases solubility of proteins.	( )
63	Dry foods are more susceptible to the browning than intermediate moisture foods.	( )
64	Dry sausages are cooked and smoked whereas semi dry sausages are not cooked.	( )
65	Emulsion type sausages have less acceptability than other type sausages.	( )
66	An air pressure of 125 psi is used with most of the stuffers.	( )
67	The term still or stationary is sometimes used to define the agitating retorts.	( )
68	Slow cooling of canned meats after heat process should be practiced to avoid quality deterioration.	( )
69	Beef stew should not contain less than 25% meat on fresh meat weight basis.	( )
70	Baked hams must attain an internal temperature of at least 140^{0}F.	( )
71	Cured meats are more resistant to discolouration than fresh meats when exposed to light.	( )
72	Nitric oxide myoglobin imparts red colour to the meat.	( )
73	The conversion of nitric oxide myoglobin to metmyoglobin is a desirable process.	( )
74	Food dehydration may be defined as the removal of water from meat during which the calorific value of food may changed.	( )
75	All dehydrated foods are subjected to two main causes of spoilage i.e. moisture and oxygen.	( )
76	Light accelerate the development of rancidity by means of oxidative spoilage.	( )
77	Meat food products advisory committee is constituted by Bureau of Indian Standards.	( )
78	All parts of meat food factory need not be disinfected.	( )
79	Galvanized iron vessels can be used to handle meat products.	( )
80	Canned meat is spoiled generally by mesophilic bacteria.	( )
81	At least 500 ppm nitrite is required in meat products to prevent germination of *Clostridium botulinum*.	( )
82	Salmonella count of meat product should be around 10 for acceptability.	( )
83	Maximum water content of sausages is not controlled.	( )

84	Cured colour in cured irradiated ham containing 25 mg/kg of sodium nitrite is not stable subsequent storage.	( )
85	Microwave processing is useful in destruction of FMD virus in meat without affecting much its flavour.	( )
86	Storage life of fresh meat can be extended up to 45 days by irradiation pasteurization technique.	( )
87	Federal regulations permit 2% or more of ice or added water in processing semi dry fermented sausage.	( )
88	Frozen flaked meat is tempered to 26-28^{0}F and then formed into desired shape using hydraulic heat press at a pressure of 500psi.	( )
89	Combination of sodium chloride and potassium chloride at a ratio of 1:1 is better than ordinary salt in meat curing.	( )
90	Dark cutting of beef is due to low glycogen reserves and low reducing sugar content.	( )
91	Nitrite level in beef rolls should not be more than 200ppm.	( )
92	For closing salamis 2 ply thread is commonly used.	( )
93	Temperature lower than 36^{0}F delay salt penetration and require longer time for both curing and salt equalization while carrying out country ham curing.	( )
94	When thermal exhausting is used during canning operation, cans are filled at 160^{0}F and exhausted for 7-10 minutes to obtain a quality product.	( )
95	Cans should be spray washed at 150^{0}F water according to meat inspection regulations.	( )
96	As per federal meat inspection regulation total amount of water in luncheon meat should not exceed 3% of the ingredients and its presence must be declared.	( )
97	Meat emulsion is a true emulsion.	( )
98	Recognition test is generally required in selection of scientific sensory panel.	( )
99	Skim milk is used as binder in comminuted meat products.	( )
100	Semidry and fermented sausages are good examples of Intermediate Moisture Meat Products.	( )

Q.4 Match the following

A.

A		B		
1.	Peelability	a.	Hypoxanthene	( )
2.	Thuringer	b.	High reflectance	( )
3.	DFD	c.	Hydrogen bond	( )

4.	Emulsifying capacity	d.	Brown discolouration	( )
5.	Meat flavour	e.	Resins	( )
6.	PSE	f.	Corn syrup solid	( )
7.	Bound water	g.	Polyphosphates	( )
8.	Cook loss	h.	Nacl	( )
9.	Maillard reaction	i.	Semidry sausages	( )
10.	Bloom	j.	High ultimate pH	( )

B.

A		B		
1.	Short meat	a.	Natural casing	( )
2.	Mixed cure	b.	Surface greening	( )
3.	Binding	c.	Hot cure	( )
4.	Cellulosic	d.	Lost binding qualities	( )
5.	Bungs	e.	Ripening	( )
6.	Tow	f.	Nitrate and nitrite	( )
7.	Nitrite burn	g.	Reduce cook out	( )
8.	Phosphates	h.	Bundle of filaments	( )
9.	Stitch pumping	i.	Cotton linters	( )
10.	Green rooms	j.	Ability to emulsify fats and water	( )

C.

A		B		
1.	Greening	a.	Light intensity	( )
2.	Colour fading	b.	Case harding	( )
3.	Rancidity	c.	Catalase	( )
4.	Dehydration	d.	Flavour	( )
5.	Polyphosphate	e.	Heat labile	( )
6.	Soy sausages	f.	High biological value	( )
7.	Water activity	g.	Deficient in meat	( )
8.	Thiamine	h.	Intermediate moisture foods	( )
9.	Meat protein	i.	Meat analogues	( )
10.	Vitamin-C	j.	pH	( )

D.

A		B		
1.	Glycogen	a.	Stabilize the meat colour	( )
2.	Pure salt	b.	Tripolyphosphate	( )
3.	Nitrite	c.	Haemoglobin	( )
4.	WHC	d.	Properties of meat products	( )
5.	Meat colour	e.	Flavour and colour problem in curing	( )
6.	Metmyoglobin	f.	Bacteriostatic	( )
7.	Cured meat	g.	Hydrostatic cooker	( )
8.	Phenols	h.	Sublimation	( )
9.	Canning	i.	Nitrosomyoglobin	( )
10.	Freeze drying	j.	Brown red colour	( )

E.

A		B		
1.	Intermediate moisture meats	a.	Curing adjuncts	( )
2.	Vapour phase	b.	Greening	( )
3.	Sorbates	c.	Antimycotics	( )
4.	Sausage	d.	Flavour of smoked meat	( )
5.	Artery pumping	e.	Antioxidant	( )
6.	Smoking	f.	Gaseous heating medium	( )
7.	Spray drying	g.	Curing	( )
8.	Sulphuring	h.	Freeze dried food	( )
9.	Space travel	i.	Heat exchangers	( )
10.	Aseptic process	j.	Browning/darkening	( )

F.

A		B		
1.	Nisin	a.	Antioxidant	( )
2.	BHA	b.	*Lactococcus lactis*	( )
3.	Redurization	c.	Symbol of irradiation	( )
4.	Redura	d.	1-5 kGy	( )
5.	Humectant	e.	$-1.5^{0}C$	( )
6.	Dried meat	f.	-20 to $-40^{0}C$	( )
7.	Freeze drying	g.	LAB	( )
8.	Start of meat freezing	h.	Glycol	( )
9.	Blast freezing	i.	Sublimation	( )
10.	Fermentation	j.	Final 4% moisture	( )

Key

Processed Meat Technology

Q.No. 1			
1.d	27.d	53.a	79.c
2.c	28.c	54.a	80.c
3.a	29.c	55.b	81.d
4.c	30.a	56.c	82.a
5.a	31.a	57.c	83.d
6.a	32.b	58.c	84.a
7.d	33.c	59.c	85.b
8.c	34.a	60.d	
9.d	35.c	61.c	
10.b	36.d	62.c	
11.d	37.b	63.d	
12.b	38.a	64.c	
13.d	39.c	65.d	
14.d	40.c	66.c	
15.b	41.a	67.c	
16.b	42.a	68.c	
17.c	43.b	69.c	
18.c	44.a	70.a	
19.c	45.b	71.c	
20.c	46.c	72.c	
21.b	47.a	73.a	
22.c	48.a	74.d	
23.d	49.b	75.c	
24.c	50.b	76.d	
25.b	51.c	77.a	
26.c	52.c	78.c	

Q.No. 2			
1.	Bologna	32.	High
2.	15^{0}C	33.	Comminution
3.	Frankfurter	34.	Nylon or surlyn
4.	Shrivelled	35.	Perishable keep under refrigeration
5.	82^{0}C	36.	15%
6.	Tangy	37.	Mortadella
7.	555 nm	38.	% transmission of light
8.	10%	39.	4.8-5.0
9.	8%	40.	3 months
10.	68.5^{0}C	41.	Fat oxidation
11.	Oxymyoglobin	42.	5.4-5.7
12.	0.5%_	43.	Component blending
13.	Free sugar amino acids	44.	Equilibrium relative humidity
14.	Mesophase	45.	Sulphur
15.	Gel	46.	Reheating
16.	Nitrosyl hemochrome	47.	Protein
17.	Protein	48.	Carbohydrate
18.	Sarcolemma	49.	Water in oil
19.	Emulsifiers	50.	High
20.	Emulsion (protein)	51.	Natural
21.	Desorption	52.	Lower
22.	Melanoidin or melanin	53.	Beginning
23.	Sulphur	54.	Binding
24.	Oriented	55.	Actin and myosin
25.	Research guiding panel	56.	Non-emulsion
26.	N-nitrosopyrolidine	57.	Meat curries
27.	%actual moisture -4x%actual protein content	58.	Cooking
28.	pH drop	59.	Nose and mouth
29.	Directly	60.	Resinous compounds and oozing out of fat.
30.	Nitrite	61.	20-25^{0}C
31.	Emulsion curing	62.	Enamal (SR)

63.	PA3679	97.	Salt
64.	1-2% and 3-5%	98.	Pale soft and exudative
65.	Clostridium botulinum	99.	Fat is emulsified
66.	Nitrous acid	100.	40-50%
67.	Flavour	101.	25-30%
68.	200ppm	102.	Air velocity
69.	Ascorbates	103.	45-50%
70.	Greater	104.	Glucose
71.	Femoral artery	105.	Minimum
72.	65-70^0	106.	50%
73.	8-10%	107.	Non-emulsion
74.	Bactericidal	108.	35%
75.	Benzpyrene	109.	Cerevalat
76.	Pyrolysis	110.	Pepperoni
77.	Stabilizes	111.	0.1-0.2%
78.	38-40^{0}F and 70-80%	112.	Pseudomonas methylotropha
79.	Salt soluble proteins	113.	Bacteria
80.	Moist heat or combination methods	114.	Filamentous fungi
81.	Moisture-protein	115.	Patis
82.	20%	116.	Bagoong
83.	Better	117.	Kind
84.	70-80%	118.	Proteins
85.	60-70%	119.	Acids and amines
86.	Tangy	120.	Lactobacillus bulgaricus
87.	Salt	121.	Butter fat
88.	40%	122.	Abnormal
89.	150-155^{0}F	123.	Keeping
90.	5	124.	Refrigeration
91.	Bound	125.	Flavour
92.	*Lactobacillus viridiscens*	126.	Sulphur
93.	*Streptococcus faecalis*	127.	Low
94.	Low	128.	PA3679
95.	Ascorbic acid and phosphates	129.	Should not
96.	4.5	130.	Nitrosamines

131.	10,000/g	139.	1.32-2.5 cm
132.	0.2g	140.	Ragioolfactoria
133.	Month and year	141.	Less than 40%
134.	450 ppm	142.	40-60%
135.	1000 ppm	143.	0.98 and above
136.	HACCP	144.	0.85-0.93
137.	Clostridium botulinum	145.	0.60-0.85
138.	100^{0}C		

Q.No. 3					
1. F	30. F	59. F	88. T		
2. T	31. T	60. F	89.T		
3. T	32. F	61. F	90. T		
4. T	33. F	62. T	91. F		
5. T	34. T	63. T	92. F		
6. F	35. F	64. F	93. T		
7. T	36. T	65. F	94. T		
8. T	37. F	66. T	95. F		
9. T	38. F	67. F	96. T		
10. F	39. F	68. F	97. F		
11. T	40. F	69. T	98. T		
12. F	41. T	70. F	99. T		
13. T	42. F	71. F	100. T		
14. T	43. F	72. T			
15. T	44. T	73. F			
16. T	45. T	74. F			
17. T	46. F	75. T			
18. F	47. F	76. T			
19. F	48. T	77. F			
20. T	49. F	78. F			
21. F	50. F	79. F			
22. T	51. F	80. F			
23. F	52. T	81. F			
24. T	53. T	82. F			
25. T	54. F	83. F			
26. F	55. F	84. T			
27. T	56. T	85. T			
28. T	57. T	86. F			
29. F	58. F	87. F			

Q.No. 4											
A		B		C		D		E		F	
1.	f	1.	d	1.	c	1	d	1.	c	1.	b
2.	i	2.	f	2.	a	2.	e	2.	d	2.	a
3.	j	3.	j	3.	d	3.	a	3.	a	3.	d
4.	h	4.	i	4.	b	4.	b	4.	b	4.	c
5.	a	5:	a	5.	j	5.	c	5.	g	5.	h
6.	b	6.	h	6.	i	6.	j	6.	e	6.	j
7.	c	7.	b	7.	h	7.	i	7.	f	7.	i
8.	g	8.	g	8.	e	8.	f	8.	j	8.	e
9.	d	9.	c	9.	f	9.	g	9.	h	9.	f
10.	e	10.	e	10.	g	10.	h	10.	i	10.	g

5

Food Packaging

Q.1 Multiple type questions

1. Silica is most commonly used in the processing of plastic packaging materials as

   a. Antiblock agent
   b. Antioxidant
   c. Antistatic agent
   d. Lubricant

2. For packaging precooked frozen meat products such as meat stews and roast or boiled sliced meats, an ideal packaging material is boilable bag consisting of

   a. Polyester film/MDPE laminates
   b. LDPE
   c. HDPE
   d. Polystyrene

3. A retortable flexible package is mainly composed of

   a. Polystyrene film/Aluminium foil/polyolefin laminate
   b. Cellophane/LDPE/Paper laminate
   c. Cellophane/HDPE/LDPE laminate
   d. LDPE/MDPE/Polystyrènes laminate

4. For bulk packaging of butter in India, containers used are made up of

   a. LDPE
   b. Aluminium foil
   c. Tinplate
   d. Polystyrene

5. Which of the following plastic film has lowest oxygen and carbon dioxide permeability

   a. LDPE
   b. Polyvinyl chloride
   c. Polyester
   d. Polyvinylidine chloride

6. 'Cryovac'is one of the method of

   a. Controlled atmosphere packaging
   b. Vacuum packaging
   c. Aseptic packaging
   d. packaging liquid milk

7. Extract of Irish moss can be used in the packaging of poultry as
   a. Plasticiser
   b. Lubricant
   c. Edible coating
   d. Antioxidant
8. Sanitary protection means protection against
   a. Air and light
   b. Micro-organisms and dirt
   c. Odour and flavour absortion
   d. Fading and discolouration
9. A food package is a structure prepared to contain a food product in order
   a. To increase the market share of the product
   b. To attract the attention of the customers
   c. To instruct the purchaser on how to use the product
   d. To protect product against contamination and loss
10. "Flash-18" process is defined as the process in which
    a. Product is filled , sealed and then processed
    b. Sterilized product is filled into sterilized container
    c. Sterilized product at 255^{0}F is filled into cleaned container which is then sealed and hold at atmospheric pressure
    d. None of the above
11. Chipboards are mainly made from
    a. Recycled waste pulp
    b. Chemical and mechanical pulp
    c. Bleached chemical pulp
    d. Resins
12. Low density polyethylene is
    a. Soluble in all solvents
    b. Insoluble in all solvents
    c. Soluble in some solvents
    d. Insoluble in some solvents
13. Packaging material that has excellent grease resistance is
    a. LDPE
    b. HDPE
    c. Polypropylene
    d. None of the above
14. Oxygen permeability is high in
    a. LDPE
    b. HDPE
    c. Polypropylene
    d. Polyvinyl chloride

15. Most commonly used type of can in meat industry
    a. Open top cans
    b. Key opening cans with reclosure
    c. Key opening cans with no closure
    d. Slip lid cans
16. Packaging graphics provide
    a. Nature of packaging material
    b. Better presentation and display
    c. Nature of the product
    d. None of the above
17. Seaming operation is important because
    a. To prevent recontamination with micro-organisms
    b. To prevent moisture losses
    c. To prevent food losses
    d. None of the above
18. Water vapour resistance is more in
    a. PE
    b. PVC
    c. PP
    d. Aluminium foil
19. Gas transmission resistance is high in
    a. PE
    b. PP
    c. PVC
    d. Kraft paper
20. Cans that are commonly used for meat products
    a. S.R. tin cans
    b. Aluminium cans
    c. Mild steel cans
    d. Alloy cans
21. Co_2 concentration of modified atmosphere for meat should range from
    a. 0-10%
    b. 10-20%
    c. Greater than 20%
    d. None of the above
22. Which of the following is the disadvantage of modified atmosphere packaging
    a. Reduced economic loss
    b. Extended transit time
    c. High quality maintenance
    d. Variable product requirement
23. Co_2 used in modified atmosphere packaging to
    a. Increases the lag phase of microorganism
    b. Does not change it any phase of microorganism
    c. Decreases the lag phase of microorganism
    d. None of the above

24. The most commonly used film for fresh meat vacuum packages is
    a. Polyethylene
    b. Polystyrene
    c. Polyvinylidine chloride
    d. None of the above
25. Fresh sausages are mostly sold in
    a. Cellulosic casings
    b. Natural casings
    c. Alignate formulations based casings
    d. None of the above
26. Best conditions for storage of eggs in the shelf are
    a. 85-92%RH and 29-32°F temperature
    b. 55-60%RH and 10-20°C temperature
    c. 70-75%RH and 2-5°C temperature
    d. None of the above
27. Which of the following milk vitamins requires special considerations for protection against sunlight
    a. Niacin
    b. Thiamin
    c. Pantothenic acid
    d. Ribiflavin
28. The expected shelf life of fresh meat even with packaging and refrigeration would be
    a. 4-6 days
    b. 4-6 hours
    c. 2-3 days
    d. None of the above
29. Indicate the true fact about plain cellulose
    a. Transparent
    b. Glossy
    c. Biodegradable
    d. All of the above
30. Natural polymer of cellulose, cellophane is widely used in packaging of ____________ when coated with nitrocellulose on one side
    a. Fresh meat
    b. Dried meat
    c. Fermented meat
    d. None of the above
31. Thickness of aluminium foil commonly used for packaging of food is
    a. 0.025-0.15 mm
    b. 0.20-0.35 mm
    c. 0.31-0.451 mm
    d. None of the above

32. The density of commonly used LDPE for food packaging is
    a. 0.123-0.228
    b. 0.56-0.625
    c. 0.91-0.925
    d. None of the above
33. Indicate the true fact about Polypropylene
    a. High strength
    b. Glossy
    c. Puncture resistant
    d. All of the above
34. Common use of polyvinyl chloride (PVC) is
    a. Pickle packaging
    b. Khoa packaging
    c. Yoghurt packaging
    d. All of the above
35. Indicate the true fact about Nylon
    a. Good mechanical strength
    b. Heat resistant (60-200°C)
    c. Clear film
    d. All of the above
36. Laminated films are most suitable for
    a. Snack food
    b. Bakery products
    c. Dried soup
    d. All of the above
37. Basic requirements for cured meat packaging is
    a. Oxygen barrier
    b. Water vapour barrier
    c. Hermetically sealable
    d. All of the above
38. Glassine bags are suitable packaging material for hamburger and hot dogs and have special quality like
    a. Microwave heating
    b. Gas barrier
    c. Heat sealable
    d. Transparent
39. Oxygen scavengers are essential in the meat of
    a. Pork and products
    b. Poultry and products
    c. Both of the above
    d. None of the above
40. Compounds used as carbon dioxide scavengers and emitters are
    a. Calcium hydroxide
    b. Calcium oxide
    c. Silica gel
    d. All of the above

41. Microsphere powder is able to release_________and acts as broad spectrum antibacterial

    a. Chlorine dioxide
    b. Sodium phosphate
    c. Potassium dichromate
    d. All of the above

42. Material used for edible packaging must be

    a. GRAS
    b. Optical property
    c. Mechanical property
    d. All of the above

43. Indicate the compound responsible for indication of spoilage and off-flavour

    a. Carbon dioxide
    b. Hydrogen sulphides
    c. Both of the above
    d. All of the above

44. Per capita consumption of packaging material in India is

    a. 4.3kg/person/annum
    b. 6.2 kg/person/annum
    c. 5.5 kg/person/annum
    d. 3.0 kg/person/annum

45. Shelf life of ready-to-serve pizza increased ___________by MAP

    a. Up to 15 days
    b. Up to 60 days
    c. Up to 45 days
    d. None of the above

46. Commercially available oxygen scavenger with trade name is

    a. Ageless
    b. Fresilizer
    c. Oxyguard
    d. All of the above

47. Beta-galactosidase (lactase) can be covalently attached to surface-functionalized LDPE and act as active packaging materials to reduce the conditions like

    a. Lactose intolerance
    b. Casein intolerance
    c. Both of the above
    d. None of the above

48. Chitosan coating films may act as _________film when used in packaging of emmental cheese

    a. Antibacterial
    b. Antioxidant
    c. Oxygen scavenger
    d. None of the above

49. Whole milk powder with multilayer active packaging film containing α-tocopherol can

    a. Destroy microbes
    b. Biosensor
    c. Delayed lipid oxidation
    d. None of the above

50. Polylactic acid plastic (PLA) is biodegradable thermoplastic derived from lactic acid is an example of

    a. Active packaging
    b. Eco-friendly packaging
    c. Edible packaging
    d. Intelligent packaging

51. Examples of thermo-formed packaging are

    a. PVC
    b. High impact poly styrene (HIP)
    c. HDPE
    d. All of the above

52. Coextruded multilayer films are

    a. LLDPE
    b. LDPE
    c. Nylon or EVOH
    d. All of the above

53. QTS can are

    a. CMQ steel
    b. Aluminium cans
    c. Grade'K' tin plates
    d. All of the above

54. Environmental factor affecting shelf life of packaged food is

    a. Light
    b. Temperature& Humidity
    c. Oxygen
    d. All of the above

55. A mixture of __________replacing air in MAP maintains red colour of meat for 15 days.

    a. 80%O_2 and 20% Co_2
    b. 50%O_2 and 50% Co_2
    c. 20%O_2 and 80% Co_2
    d. All of the above

56. Souring of cured meats on storage is caused by

    a. Streptococcus
    b. Micrococcus
    c. Both of the above
    d. None of the above

57. Dehydrated meats require protection from

    a. Oxygen
    b. Moisture
    c. Light
    d. All of the above

58. Indicate the method used for measuring the moisture protection qualities of package

    a. Weight gain loss
    b. Half value period
    c. Break down
    d. All of the above

59. Loss of moisture in frozen packaged meat products should be reduced to avoid

a. Desiccation
b. Frozen burn
c. Both of the above
d. None of the above

60. Important packaging considerations in poultry products are

a. Oxidative rancidity
b. Desiccation
c. Microbial spoilage
d. All of the above

61. MSAT cellophane means

a. Moisture proof
b. Sealable
c. Anchored and transparent
d. All of the above

62. Important packaging considerations in poultry products are

a. Oxidative rancidity
b. Desiccation
c. Microbial spoilage
d. All of the above

63. Packaging should protect the milk from contamination by

a. Bacteria/dirt
b. Light
c. O_2
d. All of the above

64. ___________perform the functions as containing, protection and to sell of the product

a. Packaging
b. Preservation
c. Processing
d. All of the above

65. Indicate the quality of aseptic packaging

a. Sterility
b. Food quality
c. Processing eligibility
d. All of the above

Q.2 Fill in the blanks

1. _______________ are added to make the packaging material more flexible and brittle.
2. In a vacuum package negative pressure on the inside and _______________ pressure on the outside push the film against the product in all directions.
3. Packaging material should provide protection against _______________ for packaging dried whole milk powder and egg powder.
4. _______________ is a two-way process which continues until the redox potentials in two phases are equal.
5. A package should be planned and designed to suit a particular product and particular _______________.
6. Water vapour permeability is defined as_______________ at 38^{0}C and 80%RH .
7. Packaging used for fresh meat should be permeable to_______________ to enhance the colour.
8. Gas packing with _______________ gives better colour in meat than N_2.
9. Bulk packaging of meat is done in _______________ boxes.
10. Egg powder is highly hygroscopic and labile to _______________.
11. Packaged cream is highly susceptible to _______________.
12. Butter must be protected from loss of __________ which causes change in colour.
13. Alluminium foil/different types of paper laminates provide opacity and good ____________ property required for packaging of butter.
14. Most natural cheeses require O_2 permeability no more than ____________/ at 73^{0}F and 50% RH.
15. The old paper board or pulp made trays used for packaging cut-up poultry birds have been replaced by the trays made of_____________________________.
16. For making hen suitable size of tray is____________________.
17. Packaging of large whole frozen poultry is usually done in copolymer film having very high____________________.
18. In blast freezer and air still freezers birds can be _______________ before freezing.
19. Lard is normally packed in_____________________________.

20. Coatings used in the inner of the cans for packing meat are called ________.
21. Meat pies are deteriorated in damp atmosphere and therefore __________ films should be used for packing them.
22. To delay oxidative deterioration of meat at room temperature, __________ is used.
23. Rigid polyvinyl chloride is a new material used for packing ____________.
24. Rigid package is advantageous because it offeres __________________.
25. The most commonly used film for vacuum packages is______________.
26. Packaging requirements of sausages are similar to _____________.
27. Fresh sausages are mostly sold in __________________.
28. Frozen meat is prone to dehydration and loss of surface texture, therefore, its packing should protect it from__________________________.
29. Packaging ______________gap between consumer and manufacturer.
30. 22 gauge tin plate is ______________ than 28 gauge tinplate.
31. 100 gauge polyethylenes is ______________than 200 gauge polyethylenes.
32. Polyethylene bags ______________ sterilized by autoclaving at 110^{0}C.
33. ______________ gives blue colour in glass.
34. Blister packaging was invented in __________________.
35. One mil=________________" inches.
36. Thickness of film is ______________ or less.
37. 400 gauge cellophane means________________.
38. _____________ slugs are used for collapsible tubes.
39. Copolymerization of PVDC with PVC makes it more plastic but reduces ____________ point.
40. ______________ film is mainly used as shrink wrap.
41. ________________is a breathing film.
42. _________________ is a linear polymer made by condensation of ethyleneglycol and terephthalate.
43. Polyethylene has low resistance to ______________________.
44. X stands for _____________ coated in cellophane films.
45. Packaging material used for fresh meat should be permeable to __________ to enhance colour.

46. Laminates are used when required properties are not achieved in a __________ packaging material.
47. Heat, wax, hot melting and ______________ are four methods of lamination.
48. ______________ test is used to evaluate all plastic materials intended for food contact application.
49. 300 gauge polyethylene film is ___________ than 400 gauge polyethylene film.
50. AFD products are_______________________.
51. Gas packaging with _______________ give better colour in meat than N_2.
52. Packaging material to be used in meat packaging should be capable of resisting _________________.
53. Control of _________________ requires a compromise between development of ideal colour and prevention of oxidative degradative reaction in red meats.
54. ________________ of fresh meat aids in tenderization and better bloom on storage.
55. The first transparent film used to wrap fresh sausages was ___________.
56. Irradiated meats are best packed in ____________________ for preventing recontaminations.
57. Most fresh meat products are placed in ___________________ with a transparent film.
58. Packaging material made of _________________ has good gloss, high flexibility and strength.
59. Polymerized vinyl chloride needs to be _____________ to make it flexible for packing oils and pickles.
60. HDPE is about_______________ less permeable to water vapour than LDPE.
61. __________________ trays offer an aesthetically appealing white background for red meats and have a quality feel.
62. __________________ was the first non-cellulosic plastic film used by meat packagers.
63. Polypropylene laminated with other films is used for ________________ and ________________ of processed meats.
64. Exposure of meat to atmospheric air for several days results in the formation of ____________________________.

65. The desired pink colour of cured meat product is attributed to the pigment ______________.

66. ______________ causes green cores in the interior of the cured meat product on storage.

67. ______________is a binding agent in between two films and has better seal property.

68. The general practice in vacuum packaging is to have a vacuum level of ______________ inches.

69. The residual level of oxygen in the controlled atmosphere is ___________.

70. Sulphur resistant internal lacquires are advocated for canned meats and meat products to prevent______________________ due to decomposition of proteins.

71. Storage of surplus food was done in________________ in mesolithic period.

72. InUSA______________ was introduced as packaging material for the first time.

73. Polyvinylidine chloride having very low permeability to water vapour and gases is being marketed as ________________.

74. In the Neolithic period ________________ were developed for storing food.

75. Parchment paper having a bursting strength of ____________ is suitable for poultry packaging.

76. In heat shrinkable films dipping the bags in hot water results____________.

77. ______________ is considered as serious problem involved in utilizing mechanically deboned poultry meat.

78. Shell eggs are conveniently stored and transported in _____________.

79. Vacuum packaging and Nitrogen gas packaging are effective packaging methods of ________________________________.

80. Polypropylene is made up of ____________________.

81. In recent years, ______________trays are commonly used for packaging of red meat.

82. ____________ films are used for packing larger and irregular cuts of fresh meat.

83. The most commonly used film for fresh meat vacuum packages is ____________.

84. ________________boxes are used for wholesale shipments of wholesale meat cuts.

85. ________________film is commonly used for frozen meat.

86. ________________films are useful for light protection.

87. ________________pouches were recently developed in place of cans for thermal processing of meat products.

88. A tight and attractive package used for dressed poultry is________________ film.

89. ________________overlap is provided on the open top of the paper board egg carton, which enables the buyer to purchase eggs.

90. Cleaned giblets are wrapped in ________________ and replaced in the body cavity.

91. The capacity of the goods to absorb moisture is one of the principal factors which affect the life of ________________.

92. The goods can tolerate at least a small change in moisture content before they begin to deteriorate and this level of moisture is known as ________________.

93. Trays used to pack cut up parts of birds are made from ________________.

94. Tetra packs used for packaging milk are ________________shaped.

95. Skim milk powder is not as much sensitive to oxidation as full fat milk powder and therefore does not require________________.

Q.3. Write true or false against each statement

1.	A package can only maintain, not improve the quality of product that goes into it.	( )
2.	The uncoated side of cellophane (coated with nitrocellulose on one side) is allowed to be in contact with meat while wrapping fresh meat for display.	( )
3.	The overwrapping film should have excellent optical properties.	( )
4.	Metal foil-plastic film laminates are widely used for the packaging of dehydrated meat products.	( )
5.	The high density polyethylene films have many more applications for packaging meats than do the low density polyethylene films.	( )
6.	Most meat packaging films are thermoplastic and extruded from the molten stage through dies.	( )

7.	Metal foil-plastic film laminates are widely used for the packaging of dehydrated meat products.	( )
8.	For packaging of cured meat products, the packaging material should have good oxygen and water vapour permeability.	( )
9.	Saran film is a copolymer of vinylidene chloride and vinyl chloride.	( )
10.	The functional characteristics of laminate do not depend upon the properties of individual packaging materials and the bonding agent.	( )
11.	By primary container we mean an outer box, case or wrapper that may hold cans or jars but does not directly contact the food.	( )
12.	PVDC coated plastic films are not suitable for cheese packaging as they provide good oxygen and moisture barrier properties.	( )
13.	As most plastics have high fat resistance, plastic films are recommended for packaging of butter for long term storage.	( )
14.	Multilayer co-extruded films LDPE/HDPE are being increasingly used for short term storage of ghee.	( )
15.	The term hermetically sealed container means a container which is absolutely impermeable to gasses and vapour throughout its entirety including its seems.	( )
16.	Polyethylene bags can be used for retort processing.	( )
17.	Packaging material for dehydrated meats should give sufficient protection against crumbling.	( )
18.	Film or foil laminates are generally used as wrap for cured meats.	( )
19.	Blotters are not necessary for packaging of fresh poultry using trays.	( )
20.	Cryovac is used for vacuum packaging of meat and poultry.	( )
21.	Saran is used for packaging of frozen poultry.	( )
22.	Coated cellophanes are used for packaging of cured meats.	( )
23.	Water vapour transmission rate increased with high humidity.	( )
24.	Cellophane is tougher when moist.	( )
25.	Meat pickles can be packed in polyethylene bags.	( )
26.	PVDC is suitable for hygroscopic products.	( )
27.	Polystyrene is suitable for packaging fresh meat and poultry.	( )
28.	MR or MC steel base plates are used for making the cans for packing mildly corrosive foods.	( )
29.	Apricot and figs are classified as moderately corrosive foods and MR type steel base plates are most suitable for making cans for these foods.	( )

30.	Nitric oxide myoglobin imparts red colour to the meat.	( )
31.	The book "Food Packaging" has been written by Stanley Sacharow and R.C.Griffin.	( )
32.	Water vapour transmission rate increased with high humidity.	( )
33.	Cellophane is tougher when moist.	( )
34.	Low density polyethylene is a good gas barrier.	( )
35.	Oil pickles can be packed in polyethylene bags.	( )
36.	P.T. cellophanes are a good barrier against non-boring insects.	( )
37.	P.T. cellulose film is sealed by heat sealing.	( )
38.	PVDC is suitable for hygroscopic products.	( )
39.	Polystyrene is suitable for packaging fresh meat and poultry.	( )
40.	Wider heat seal ranges are desirable.	( )
41.	Polyester does dissolve in benzoyl alcohol.	( )
42.	The uncoated side of cellophane (coated with nitrocellulose on one side) is allowed to be in contact with meat while wrapping fresh meat for display.	( )
43.	Metal foil-plastic film laminates are widely used for the packaging of dehydrated meat products.	( )
44.	PVDC coated plastic film are not suitable for cheese packaging as they provide good oxygen and moisture barrier properties.	( )
45.	The HDPE film have many more applications for packaging meats than do the LDPE films.	( )
46.	Saran film is a copolymer of vinylidene chloride and vinyl chloride.	( )
47.	N_2 gas packaging of egg powder gives better results.	( )
48.	Corrugated board has no cushioning properties.	( )
49.	Packaging of fresh meat is intended to prevent bacterial contaminations.	( )
50.	Cellophane coated with nitrocellulose on one side is permeable to oxygen but impermeable to water vapour.	( )
51.	Heat shrinkable films producing tighter packaging are no way better than nonshrink films.	( )
52.	Rubber hydrochloride is more durable than cellophane.	( )
53.	LDPE film of 150 gauge is permeable to oxygen and provides sufficient moisture vapour barrier.	( )
54.	Polyethylene is widely used because of its extensibility, low strength clarity and gloss.	( )
55.	Skinless sausages are generally linked in cellophane casings and are pealed off after cooking and before packaging.	( )
56.	Polyethylene in the perforated form has been used to some extent for packing fresh sausages.	( )

57.	At present shrink wrapping accounts for 25% of the fresh meat marketed in packets.	( )
58.	Vacuum packaging provides oxygen and moisture barrier for meats and is suitable for storage upto 3 weeks.	( )
59.	Commercially sterile canned meats are shelf stable for a maximum period of one year room temperature.	( )
60.	Vacuum packed sliced bacon could be kept in good condition for 6-8 months at 0-4^{0}C.	( )
61.	Thermoprocessed meat products packed in retort pouches are shelf stable for a period of over one year.	( )
62.	Controlled atmospheric packaging is an alternative to vacuum packaging.	( )
63.	In modified atmospheric packaging, beef needs high nitrogen and carbon dioxide content in the pack to maintain bright red colour.	( )
64.	Inspite of some loss in colour of meat, 100%carbondioxide is beneficial in maintaining microbial quality.	( )
65.	A decrease in pH on storage contributes to the oxidation of nitrosomyoglobin to metmyoglobin in cured meats.	( )
66.	Controlled atmospheric packs have very good performance in the freezer.	( )
67.	Polyethylene obtained by exposing ethylene to very high pressure and temperature is most widely used flexible packaging material.	( )
68.	The water vapour permeability rate of LDPE increases by 2-5 fold with every 10^{0}C raise in storage temperature.	( )
69.	The oxygen permeability of polyethylene falls sharply at refrigerated storage.	( )
70.	HDPE is obtained by high temperature (150-200^{0}C) and high pressure polymerization of ethylene.	( )
71.	LDPE and HDPE are polymers of ethylene subjected to different pressures and temperature.	( )
72.	LDPE is obtained by low temperature (60-150^{0}C) and low pressure (40 atmosphere) polymerization of ethylene in the presence of a metal catalyst.	( )
73.	Biaxially oriented polystyrene film has been used for meat packages because of its clarity, permeability and aesthetically pleasing presentation.	( )
74.	Flexible packaging materials are most suitable for dressed whole birds, half birds or cutup parts.	( )
75.	Wooden crates are widely used for bulk transportation of poultry from dressing units to retail outlets.	( )

76.	Egg is bestowed with natural package with shell and shell membranes and needs no extra protection.	( )
77.	Ten percent salt or sugar may be added to broken eggs to be frozen to enhance coagulation and rubbariness.	( )
78.	Vacuum packaging in shrinkable films of high tear resistance is good for frozen poultry.	( )
79.	Oxymyoglobin is well maintained in vacuum packed meat.	( )
80.	Lactobacilli are predominant in anaerobic packaging.	( )
81.	Packaging cannot prevent the freezer burn in frozen meat.	( )
82.	Cured meats are normally packed under vacuum.	( )
83.	Cooling process immediately after sterilization of canned meats helps to retard spoilage.	( )
84.	Cryovac films extend the shelf life of meat and poultry packed under aerobic condition.	( )
85.	Sulfonated films are used for precooked meats.	( )
86.	Heat shrinkable bags are well suited to preserve pork carcasses against bacterial spoilage during extended periods of shipping and storage.	( )
87.	Liquid sodium hypochlorite is the most widely used sanitizer in canned meat industry.	( )
88.	Titaniumdioxide and carbon black are added to regenerated cellulose films to reduce light transmission through the film.	( )
89.	Packaging requirements for natural cheese are less critical than processed cheese.	( )
90.	The term hermetic refers to a container that is absolutely impermeable to gases only.	( )
91.	Packaging of food is one of the methods of preservation which is termed as asepsis.	( )
92.	For ice-cream packaging protection against moisture loss and temperature fluctuations is desirable.	( )
93.	Meat and meat products do not require lacquer cans for packing.	( )
94.	MR or MS type steel base plates are used for making cans for meat packing.	( )
95.	Vacuum packing is normally undesirable for frozen meats.	( )
96.	Light fading in cured meat, can be checked by protecting the package from light.	( )
97.	The most common method of packing shelled eggs is egg cartons.	( )
98.	The use of foil plastic laminates is recommended for packing of butter.	( )

99.	Nanocomposites can be utilized to improve the barrier properties of plastic wrapping for foods and dairy products.	( )
100.	Biosensor based commercially available Food Sentinel System is based on the principle of immunological reactions and detects contamination.	( )

Q.4 Match the following

A.

A		B		
1.	Frozen poultry	a.	Butter	( )
2.	Gas flushing	b.	Liquid milk	( )
3.	Chicken curry	c.	Barrier properties	( )
4.	Aluminium foil	d.	Copolymer	( )
5.	Vacuum packaged meat	e.	Antioxidant	( )
6.	Tetra pack	f.	Plasticiser	( )
7.	Vegetable parchment paper	g.	Shrink packaging	( )
8.	Butylated hydroxy anisole	h.	Nitrogen	( )
9.	Ethylene –Vinyl Acetate (EVA)	i.	Canning	( )
10.	Phthalates	j.	*Microbacterium thermosphectum*	( )

B.

A		B		
1.	Mildly corrosive	a.	Nicholas Appert	( )
2.	Canning	b.	Alumina	( )
3.	Glass	c.	Finish	( )
4.	Glass container	d.	Pressurized gas or liquid	( )
5.	Composite container	e.	Peas and corn	( )
6.	Aerosol container	f.	Paper board body with metal or plastic ends	( )
7.	Butter	g.	Cured meat colour	( )
8.	Nitrosomyoglobin	h.	Flexible can	( )
9.	Retort pouch	i.	Parchment paper	( )
10.	Hollow plastic corrugated board	j.	Cellular profile board	( )

C.

A		B		
1.	Corrugated paper board	a.	Q.A.	( )
2.	GMP	b.	Hedonic scale	( )
3.	Subjective score	c.	Linear and fluting media	( )
4.	Sterilization of metal container	d.	Radiation	( )
5.	Sterilization of plastic containers	e.	Copolymer coated cellulose/PE film laminates	( )
6.	Sterilization of glass containers	f.	Alternative vacuum packaging	( )
7.	Sterilization of heat sensitive plastic containers	g.	Polyethylene	( )
8.	Suitable film for vacuum packaging	h.	Ethylene dioxide	( )
9.	CAP	i.	Hydrogen peroxide (H_2O_2)	( )
10.	Short term storage of cured meat	j.	Super heated steam	( )

D.

A		B		
1.	Ham and Bacon	a.	Hot dipping	( )
2.	Vacuum packaging or gas flushing	b.	Cleanliness and coolness	( )
3.	Tin plates	c.	Tranquility	( )
4.	Blue colour of packaging material	d.	Warmth	( )
5.	Yellow colour of packaging material	e.	Sumerian times	( )
6.	Green colour of packaging material	f.	Alpine tribes	( )
7.	Orange colour of packaging material	g.	Plasticization	( )
8.	PVC	h.	Strength and cheerfulness	( )
9.	Barrel for packaging	i.	Cellophane /PVDC/ Polyethylene laminates	( )
10.	Glass packaging	j.	Shrinkable PVDC/PVC copolymer	( )

E.

A		B		
1.	Paper packaging	a.	Chinese invention of 16th century	( )
2.	Cardboard packaging	b.	Invented in 1934	( )
3.	Rubber hydrochloride	c.	Replacement of parchment	( )
4.	Swollen can	d.	A canning operation	( )
5.	Retort pouch	e.	Clostridia	( )
6.	Polyethylene packaging	f.	High oxygen barrier film	( )
7.	Exhaustion	g.	Polyethylene and PVC	( )
8.	Saran	h.	Aluminium foil	( )
9.	Dual film system	i.	Aerobic microbial spoilage	( )
10.	Frozen hamburger	j.	F_0 value	( )

Food Packaging Key

Q. No. 1		Q. No. 2	
1.a	38.a	1.	Plasticisers
2.a	39.c	2.	Positive
3.a	40.d	3.	O_2 and moisture
4.c	41.a	4.	Oxidation
5.d	42.d	5.	Function
6.b	43.c	6.	gm/sq.m/24hrs
7.c	44.a	7.	Oxygen
8.b	45.c	8.	Co_2
9.d	46.d	9.	Corrugated fibre board
10.c	47.a	10.	Rancidity
11.a	48.a	11.	Loss of moisture
12.a	49.c	12.	Moisture
13.c	50.b	13.	Barrier
14.a	51.d	14.	5cc/100 rq in/24 hrs
15.a	52.d	15.	Polystyrene foam or clear polystyrene
16.b	53.d	16.	5" x10" x1"
17.a	54.d	17.	Puncture resistance
18.d	55.a	18.	Cartooned
19.c	56.c	19.	Paper wraps or waxed cartons
20.a	57.d	20.	Meat enameles
21.c	58.d	21.	Breathing
22.b	59.c	22.	Foil ply
23.b	60.d	23.	Cured meat
24.c	61.d	24.	Better stocking and increased protection
25.b	62.d	25.	Polyvinylidine chloride
26.a	63.d	26.	Fresh meat
27.d	64.a	27.	Natural casings
28.a	65.d	28.	Moisture loss and temperature fluctuations
29.d		29.	Bridges
30.a		30.	Thicker
31.a		31.	Thinner
32.c		32.	Can't
33.d		33.	COO_2
34.a		34.	1952
35.d		35.	0.01"
36.d		36.	0.254cm
37.d		37.	40gm/$10m^2$

38.	AI	67.	Ionomer
39.	SP	68.	25-30 inches
40.	PP	69.	Less than 1%
41.	PS	70.	Sulpur staining or discolouration
42.	Polyester	71.	Basket made of grass
43.	Grease/oil	72.	Cellophane
44.	PVDC	73.	Saran or cryovac
45.	O_2	74.	Pottery and metal containers
46.	Single	75.	7.5 kg/cm^2
47.	Co-extrusion	76.	Contour package of the product
48.	Global migration	77.	Lipid oxidation
49.	Thinner	78.	Moulded pulp filler flats
50.	HDPE or Polypropylene	79.	Mechanically deboned poultry meat (MDPM).
51.	CO_2	80.	Hydrocarbons
52.	Tearing and puncturing	83.	Polyvinyl chloride
53.	Oxygen permeation	84.	Corrugated
54.	Vacuum packaging	85.	LDPE
55.	Coated heat sealable cellophane	86.	Foil laminate
56.	Metal cans_or an impervious flexible laminates	87.	Retortable
57.	Trays and overwrapped	88.	PVC
58.	PP	89.	Shrink film
59.	Plasticized	90.	Polyethylene
60.	Three times	91.	Dehydrated goods
61.	Polystyrene foam	92.	Critical moisture
62.	Rubber hydrochloride	93.	Expanded polystyrene foam and polystyrene
63.	Vacuum packaging and gas packaging	94.	Tetrahedral
64.	Metmyoglobin or brown pigment	95.	Vacuum and gas packaging
65.	Nitrosomyoglobin		
66.	Lactobacillus		

Q.No. 3					
1.	T	41.	T	81.	F
2.	T	42.	T	82.	T
3.	T	43.	T	83.	T
4.	T	44.	F	84.	F
5.	F	45.	F	85.	T
6.	T	46.	T	86.	T
7.	T	47.	F	87.	T
8.	F	48.	F	88.	T
9.	T	49.	F	89.	F
10.	F	50.	T	90.	F
11.	F	51.	F	91.	T
12.	F	52.	T	92.	T
13.	F	53.	F	93.	F
14.	T	54.	F	94.	F
15.	T	55.	F	95.	F
16.	F	56.	T	96.	F
17.	T	57.	F	97.	F
18.	T	58.	T	98.	F
19.	F	59.	F	99.	T
20.	F	60.	F	100.	T
21.	F	61.	T		
22.	T	62.	T		
23.	T	63.	F		
24.	F	64.	T		
25.	F	65.	T		
26.	T	66.	F		
27.	T	67.	F		
28.	T	68.	T		
29.	T	69.	T		
30.	T	70.	F		
31.	T	71.	T		
32.	T	72.	F		
33.	F	73.	T		
34.	F	74.	T		
35.	F	75.	F		
36.	T	76.	F		
37.	T	77.	F		
38.	T	78.	F		
39.	T	79.	F		
40.	T	80.	T		

Q.No. 4									
A		B		C		D		E	
1.	g	1.	e	1.	c	1	j	1.	c
2.	h	2.	a	2.	a	2.	i	2.	a
3.	i	3.	b	3.	b	3.	a	3.	b
4.	c	4.	c	4.	j	4.	b	4.	e
5.	j	5.	f	5.	i	5.	h	5.	j
6.	b	6.	d	6.	h	6.	c	6.	i
7.	a	7.	i	7.	d	7.	d	7.	d
8.	e	8.	g	8.	e	8.	g	8.	f
9.	d	9.	h	9.	f	9.	f	9.	g
10.	f	10.	j	10.	g	10.	e	10.	h

6

Food Microbiology

Q.1 Multiple type questions

1. Enumerate the intrinsic factor affect the microbial growth
   - a. pH
   - b. Moisture
   - c. Oxidation reduction potential
   - d. All of the above
2. Enumerate the extrinsic factor affect the microbial growth
   - a. Storage temperature
   - b. Relative humidity
   - c. Presence and concentration of gas
   - d. All of the above
3. Name the method used for determination of micro-organisms in food
   - a. DMC
   - b. SPC
   - c. MPN
   - d. All of the above
4. Name the radiation important in food preservation
   - a. UV rays
   - b. Beta rays
   - c. Gamma rays
   - d. All of the above
5. Name the organism responsible for food poisoning almost solely from consumption of sea foods
   - a. *Streptococcus haemolyticus*
   - b. *Staphylococcus aureus*
   - c. *Vibrio parahaemolyticus*
   - d. All of the above
6. Computer based mathematical models are used to predict microbial effects on
   - a. Product composition
   - b. Product processing
   - c. Products shelf life
   - d. All of the above
7. Verification step in HACCP should include
   - a. Analysis of consumer complaints
   - b. Audit of HACCP plan
   - c. Audit of documentation system
   - d. All of the above

8. The basic intension of food regulations is to
   a. Prevent fraud
   b. Protect the public health
   c. All of the above
   d. None of the above
9. Frozen samples of studies on microbiological quality should be
   a. Thawed at 15°C
   b. Thawed at 0°C
   c. Thawed at 8°C
   d. Thawed at 3°C
10. Phosphorescence caused by *Pseudomonas phosphorescence*
    a. Is never visible
    b. Is visible even after decomposition
    c. Disappear as decomposition develops
    d. All of the above
11. *Lactobacillus viridiscence* produce hydrogen peroxide and oxidizes the meat pigment to produce
    a. Red colour
    b. Greyish to greenish colour
    c. Voilet colour
    d. Brownish to blackish colour
12. Pseudomonas are sensitive to
    a. Low salt concentration
    b. High sugar concentration
    c. High oxidation-redox potential
    d. All of the above
13. Pseudomonas spp. are the predominant spoilage organisms on
    a. Cured meat products
    b. Canned meat products
    c. Fresh refrigerated meat
    d. All of the above
14. Under Good Manufacturing Practices temperature of the raw meat to avoid Salmonella growth should be
    a. Never be >12°C
    b. Never be >7°C
    c. Never be considered
    d. None of the above
15. Sanitary quality of food is indicated by
    a. Salmonella count
    b. Enterococci and Coliform count
    c. Lactic acid bacteria
    d. All of the above
16. Most spoilage bacteria do not grow below a_w
    A. 0.80
    B. 0.91
    C. 0.75
    D. 0.99

17. Aerobic organisms require-----------------Eh values for growth.

A. +ve or oxidized B. –ve or reduced

C. Both of the above D. None of the above

18. The range of temperature on which micro-organisms are reported to grow

A. – 34°C to 100°C B. 0°C -100°C

C. 7°C -65°C D. None of the above

19. The lowest reported a_w value for food borne bacteria is 0.75 for

A. Halophiles B. Mesophiles

C. Thermophiles D. Psychrophils

20. Chemical reactions causing oxidative rancidity is called

A. Dehydration B. Shrinkage

C. Autooxidation D. Discolouration

21. When poultry meats undergo low-temperature spoilage then the primary spoilage organisms belong to the genus

A. Corynebacterium B. Pseudomonas

C. Chlamydia D. Staphylococcus

22. ------- is the single most important atmospheric gas used to control micro-organisms in food.

A. O_2 B. Co_2

C. Nitrogen D. None of the above

23. -------------is the bacterium most widely used as an indicator of the sanitary state of fresh foods.

A. *Escherichia coli* B. *Salmonella typhimurium*

C. *Staphylococcus aureus* D. None of the above

24. The essential feature of poultry spoilage is ________at the outer surfaces of the carcass or cuts.

A. Bitterness B. Gas formation

C. Sliminess D. Shrinkage

25. Micro-organism responsible for hay like odour in egg is

A. Enterobacter sp. B. *E.coli*

C. Pseudomonas sp. D. Achromobacter sp.

26. Detection of specific pathogenic bacteria such as Salmonella, using traditional techniques may require up to

A. 24 hrs
B. 72 hrs
C. 2 days
D. 7 days

27. ___________ is an important food safety management system in poultry processing plant.

A. ISO-9001
B. ISO-9002
C. ISO-9003
D. HACCP

28. Eradication of Salmonella from carcasses occurred with a chlorine concentration of

A. 300-400 ppm
B. 100-200 ppm
C. 500-800 ppm
D. 800-1000 ppm

29. Indicator organisms of faecal contamination is

A. PAGE
B. *Salmonella typhimurium*
C. *E. coli*
D. AGPT

30. Which one is not commonly used for detection of spoilage in meat

A. Extract release volume
B. Dye reduction test
C. Thiobarbituric acid value
D. Immunodiffusion test

31. Indicate the source of infection through utensil at the point of storage before pasteurization

A. *Staphylococcus aureus*
B. Yeast
C. Coliforms
D. All of the above

32. Organisms can survive but not growing at pasteurization temperature are

A. Thermoduric
B. Thermophilic
C. Both of the above
D. None of the above

33. Thermophilic species are spore forming, heat loving microorganism and grow best at 55°C and their presence indicate re-pasteurization or insanitary practices

A. *Bacillus* spp
B. *Lactobacillus thermophilus*
C. Both of the above
D. None of the above

34. Presence of Psychrophils ______________indicate use of polluted water for rinsing or in other operations

    A. Pseudomonas spp B. Achromobacter spp
    C. Both of the above D. None of the above

35. UHT milk(135-150°C) may be contaminated with

    A. *Bacillus stearothermophilus* B. *Lactobacillus thermophilus*
    C. Both of the above D. None of the above

36. Bacteria used for acidophilus milk production is

    A. *Lactobacillus lactis* B. *Lactobacillus thermophilus*
    C. *Lactobacillus acidophilus* D. None of the above

37. Bacteria used for bulgarian yoghurt production is

    A. *Lactobacillus lactis* B. *Lactobacillus thermophilus*
    C. *Lactobacillus delbrueckii* D. None of the above

38. BIS Bacteriological standards for good quality raw milk is

    A. DMC <5 lakhs/ml B. SPC 2-10 lakhs/ml
    C. MBRT 3-4 hrs D. All of the above

39. BIS Bacteriological standards for pasteurized milk is

    A. Coliform count-absent in 1:10 dilution
    B. Maximum SPC 30000cfu/ml
    C. MBRT >4 hrs
    D. All of the above

40. BIS Bacteriological standards for ice-cream is

    A. Coliform count-not more than 90 B. SPC not more than 2,50,000
    C. Phosphatase test-ve D. All of the above

41. BIS Bacteriological standards for Whole Milk Powder is

    A. Coliform count-absent in 0.1g B. SPC not more than 40,000
    C. Salmonella –absent in 25 g D. All of the above

42. BIS Bacteriological standards for Skim Milk Powder is

    A. Coliform count-absent in 0.1g B. SPC not more than 50,000
    C. Both of the above D. All of the above

43. Name the most common bacteria in aerobic spoilage of meat particularly responsible for dense slime formation (109/cm^2)

    A. Pseudomonas spp B. Acenitobacter spp

    C. Morexella spp D. All of the above

44. Name the bacteria causes spoilage both in aerobic and anaerobic conditions in meat

    A. *Pseudomonas fluorescence* B. *Acenitobacter putrifaciens*

    C. *Morexella indica* D. All of the above

45. Name the bacteria causes spoilage of vacuum packed meat

    A. *Pseudomonas fluorescence* B. *Acenitobacter putrifaciens*

    C. *Brocothrix thermospacta* D. All of the above

46. Indicate the bacteria which is dominating spoilage flora in meat at <4°C

    A. Pseudomonas B. Acenitobacter

    C. Morexella D. All of the above

47. Indicate the bacteria which is dominating spoilage flora in meat at and above 10°C

    A. Pseudomonas B. Acenitobacter

    C. Staphylococcus D. All of the above

48. Indicate the organisms which can survive in freeze-dried meats

    A. *Staphylococcus aureus* B. *Trichenella spiralis*

    C. Both of the above D. None of the above

49. Indicate the common contaminates of meat and poultry which are sensitive to irradiation

    A. Salmonella B. Compylobacter

    C. Listeria D. All of the above

50. Microbiological limits for fresh meat are

    A. SPC <105 B. Coliforms <102

    C. E. coli 50 D. All of the above

51. The specifications for chilled meat are

    A. SPC >106 B. No pathogen

    C. Both of the above D. All of the above

52. The specifications for frozen meat are

A. SPC 107
B. Coliforms <104
C. Both of the above
D. All of the above

53. Name the most common gases found in cans of spoiled foods

A. Hydrogen
B. Carbon-di-oxide
C. Hydrogen sulphide
D. All of the above

54. Fungal agents responsible for spoilage of eggs in the cold room in India are

A. Aspergillus
B. Penicillium
C. Cladosporium
D. All of the above

55. The low acid food spoiled by sulfide type of organisms has usually

A. Black Colour
B. Rotten egg odour
C. Both of the above
D. All of the above

56. Microorganism fails to grow in fresh albumin due to the presence of

A. Lysozyme
B. Avidin
C. Conalbumen
D. All of the above

57. Chief types of spoilage of canned foods by thermophilic spores forming bacteria are

A. Flat sour
B. TA spoilage
C. Sulfide spoilage
D. All of the above

58. Yellow milk is due to multiplication of organisms like

A. *Pseudomonas synxantha*
B. Flavobacterium
C. Both of the above
D. None of the above

59. Fish and sea foods are generally spoiled by

A. Autolysis
B. Oxidation
C. Bacterial activity
D. All of the above

60. In fresh meat spoilage, the lactic acid bacteria are responsible for producing

A. Slime
B. Discolouration
C. Souring
D. All of the above

61. Principle type of spoilage in butter caused by bacteria is
    - A. Surface taints
    - B. Putridity
    - C. Rancidity
    - D. All of the above
62. Common pathogen of cooked meat is
    - A. *Clostridium perfringens*
    - B. *Salmonella typhimurium*
    - C. *E.coli*
    - D. All of the above
63. The pH of meat is in the range of
    - A. 5.6-7.4
    - B. 4.5-8.5
    - C. 4.5-6.5
    - D. All of the above
64. Most important agents for poultry meat contamination are
    - A. Salmonella spp
    - B. Campylobacter spp
    - C. Both of the above
    - D. All of the above
65. When poultry meat is stored aerobically under chill conditions, the organisms that predominate at spoilage invariably are
    - A. Pseudomonas spp
    - B. Salmonella spp
    - C. Campylobacter spp
    - D. All of the above

Q.2 Fill in the blanks

1. Green rots of spoiled eggs are due to ____________________.
2. Serratia spp. produces ____________________.
3. Spoilage thermophile viz. *B. stearothermophilus* does not grow at ________.
4. The optimum temperature to grow for spoilage thermophile viz. *B. stearothermophilus* is around ______________.
5. A can is called ______________when one end can be made to become convex by striking or heating the can.
6. Can with both ends bulged when one or both remain concave if pushed in or when one end is pushed and the other pops out is called ____________.
7. Quick or fast freezing is the process by which the temperature of food is lowered to about ________________.
8. The time necessary to kill a given number of organisms at specific temperature is called ____________________.
9. The time required to destroy 90% of organisms is expressed as __________.

10. The organism __________________ is the only bacterium of public health importance that can grow at a_w value near 0.86.
11. Fresh iced fishes are invariably spoiled by _________________________.
12. Salted and dried fishes are spoiled by ______________________.
13. Organisms that grow in high concentration of salt are called__________.
14. Organisms that can withstand but not grow in high concentration of salt are called____________________.
15. The lowest reported temperature for growth of Staphylococci and Salmonellae in food is ________________.
16. The lowest reported temperature for growth and toxin production for Type A and B of *Clostridium botulinum* is ________________________.
17. Skin surface of dressed chicken contains amines like Putriscine, Cadavarine, histamine and tyramine are considered as indicators of ______________.
18. Lactobacilli and Pediococci assist in controlling colonization of Salmonella and Compylobacter in the gastro intestinal tract of broiler chicken through production of ________________________.
19. Lactic acid, acetic acid, hydrogen peroxide, low molecular weight nitrogen free compounds acts as ___________________ in bacteriocin production.
20. Aerobic microbial agents found in meat can grow well at water activity level ______________.
21. Aerobic microbial agents found in meat can grow well at pH ___________.
22. Salmonella bacteria of meat can grow at water activity level_____________.
23. Staphylococcus bacteria of meat can grow at water activity level ________.
24. Halophilic bacteria of meat can grow at water activity level_______________.
25. Greenish discolouration of meat is caused by salt tolerant organisms capable of growing at_____________________________.
26. Turbidity development due to microbial growth in pickled pig feet is rarely observed unless______________is added to vinegar pickle.
27. Abdominal pain/ cramps is rarely observed in food (meat) poisoning due to________________________.
28. Canned poultry is more often spoiled by putrefactive rather than Saccharolytic clostridia due to_________________________________.
29. Burnt of caramel flavour in milk is due to certain strains of organisms like ____________________________.

30. Off flavour in meat occurs when microbial load exceeds_______________.
31. Slime development in meat develops when microbial load exceeds _______________.
32. Fish should not contain more than_______________ aerobes.
33. Breaded fish should not contain more than _______________ aerobes.
34. The ultraviolet lamps in meat ageing storage should be which produce _____________ amount of ozone.
35. Aseptic canning embodies a combination of the principles of_____________ sterilization.
36. Usually the spores of food spoilage bacteria will not germinate in the acid food. The acid food pH is considered below____________pH level.
37. The protein _____________ is an example of bacteriostatic substance that naturally occurs in the food/meat.
38. Molds, yeast and mesophilic bacteria dominate the spoilage at temperature ranging from______________ ^{0}C.
39. Protein of meat is impermeable to bacteria, but the bacteria use meat for their growth through their _______________.
40. The mold is ready growers on refrigerated meat surface as they are able to grow at ____________ ^{0}C.
41. The sour hip condition in stored meat reveals _______________ due to bacteria.
42. The greenish pigmentation of meat develops due to multiplication of _______________ producing bacteria.
43. Carbon dioxide in canned food is produced by development of __________ utilizing bacteria.
44. Most common food poisoning Salmonella is ____________________.
45. Staphylococcus food poisoning is manifested after ____________ consumption of food.
46. *Clostridium botulinum* food poisoning is manifested after ____________ of food consumption.
47. The Ptomine poisoning refers to _______ resulting from the decomposition of proteins.
48. Caseous lymphadenitis is observed only in ______________ carcasses at meat inspection.
49. The primary spoilage organism of poultry meat is____________________.

50. The flora of spoiled fish is mainly consist of ____________________.
51. Microbial deterioration of eggs was first reported by ______________ in 1873.
52. Isolation of *Streptococcus lactis* in pure culture was done by __________ in 1873.
53. Study on thermophilic bacteria was first time done by ____________ in 1888.
54. First time determine the number of bacteria in milk by ________________.
55. ______________ use the term psychrophile first time for the organisms that grow at 0^0C.
56. First isolation of *Bacillus coagulans* from coagulated milk was done by ______________ in 1915.
57. First isolation of *B. steriothermophilus* from cream style corn was done by ______________.
58. First isolation of *Salmonella enteritidis* from meat which caused food poisioning was done in ________________.
59. ________________ first discovered *Clostridium botulinum* in 1896.
60. First report of production of aflatoxins by ______________________.
61. ____________________term is used to define ratio of water pressure of food substrate to the vapour pressure of pure water at the same temperature.
62. ___________________ term is used for enhancement of keeping quality of food causing substantial reduction in number of viable spoilage microbes by radiation.
63. ________________term is used for aerobic bacteria which grow better under slightly reduced conditions.
64. ________________ term is used for organisms that are able to grow in high concentration of sugar.
65. ____________________ term is used for organisms that grow well at and above 45^0C with optima between 55-65^0C.
66. ______________ is a antimicrobial constituent in milk and effective against streptococci.
67. Fresh milk is effective against coliform due to presence of ____________.
68. Egg white possesses antimicrobial property due to presence of __________.
69. Cinnamon possesses antimicrobial property due to ______________.

70. Cloves have antimicrobial property due to ______________________.

71. Sliminess or ropiness of food may be due to the presence of ____________ in bacteria.

72. Of primary interest to food microbiologists are the spore forming species genera ________and ____________.

73. The presence of ______________ in food may cause either its spoilage or discolouration or produce aflatoxin in foods which produce disease in man.

74. Putrifaction of meat type spoilage is under ______________ condition.

75. Surface slime on meats type spoilage is under ______________ condition.

76. With respect to fish spoilage, the most susceptible part of fish is _________.

77. With respect to fish spoilage, ______________are the last part of fish to be decomposed.

78. Meat was first time preserved in glass bottles by a Persian confectioner named ______________in 1809.

79. Approximate minimum a_w values for *Cl. botulinum* is ______________.

80. Approximate minimum a_w values for Aspergillus is ________________.

81. Approximate minimum a_w values for Halophilic bacteria is _________.

82. Approximate minimum a_w values for *E. Coli* is ________________.

83. Organic acids are usually ineffective against meat spoilage organisms at pH ____________.

84. Water activity of muscle tissue is about ____________and as such does not resistant bacterial growth in it.

85. In case of aerobic flora Pseudomonas maintains a growth rate advantage over competing _______________________ at 20⁰C.

86. Minimum growth temperature of molds in meat is ________________.

87. ______________ group organisms are particularly sensitive to Co_2.

88. ____________________ group of organisms are not sensitive to Co_2.

89. Enterobacteriasae can ferment substrate like glucose-6-phosphate present in meat but not ____________________.

90. Generation time (hrs) in meat at 10⁰C is less for Pseudomonas as compared to ______________________.

91. The inhibitory effect of Co_2 upon Pseudomonas and other putrifying organisms might result from inhibition of the ___________________.

92. Inhibition of all microbial growth in meat occurs when the freezing temperature reaches ________________.

93. Cooked beef roast should be heated to an internal temperature of at least ________________ to assure Salmonella destruction unit.

94. Most important food source of Salmonella is________________meat.

95. More organisms are killed between ________________________.

96. ______________________ stops multiplication of all the organisms.

97. Ionizing radiations______________the molecules in their path thus causing irreversible damage to the micro-organisms.

98. ______________________ is one of the most radio resistant bacteria known.

99. A low nitrite level of ________________ ppm is sufficient for colour and flavour development.

100. Atleast ______________ppm nitrite level is required for antibotulinal activity in the processed meat products.

101. Nisin is produced by ____________________________.

102. The mould _________________ is associated with the formation of buttons on the surface sweetened condensed milk.

103. Ropiness in peaches is caused by ______________________.

104. Outgrowth of *Clostridium botulinum* spores is prevented if the pH does not exceed______________.

105. Several hundred cases of *Eneteritis necroticans* occurred in Germany after world war-II following consumption of canned meat. The illness was caused by ________________.

106. The abbreviation SSCCM stands for ________________________.

107. ________________ is G+ve cocci producing heat stable eneterotoxins commonly in milk products .

108. _______________G-ve psychotrophs bacteria producing slime, fluorescent compounds and a variety of pigments on cold stored foods.

109. ________________ G-ve pathogen commonly found in poultry and has more than 2000 serotypes.

110. ______________________ cause souring of foods by producing acetic acid from alcohol.

111. __________________________can cause phosphorescence of meat.

112. ____________________ is eneterococci than can grow in vacuum packaged food under cold storage.

113. It has been shown that the rate of spoilage of fresh poultry at 10^0C is ____________________ then that at 5^0C

114. Yeasts preferring high osmotic pressure are called ____________________.

115. AOAC stands for ____________________.

116. ____________________ is the most widely used bacterium as an indicator of the sanitary state of fresh poultry meat.

117. The organisms required optimum temperature for growth in between 55^0C to 65^0C are termed as ____________________.

118. The group of organisms responsible for surface spoilage is called ____________________.

119. In direct microscopic count of organisms, cell counts are performed under ____________________.

120. Hurdle concept was coined by ____________________.

121. Whole poultry tends to have ____________________ microbial count than cut-up poultry.

122. ELISA stands for ____________________.

123. The term psychrophile was coined by ____________________ in 1902.

124. Metabolically injured organisms are not able to form colonies on ____________________.

125. Common pathogen of poultry meat is ____________________.

Q.3. Write true or false against each statement

1.	Limulus lysate test detects exotoxins present in the extracts of foods.	( )
2.	DNase produced by pathogen *Staphylococcus aureus* is heat labile.	( )
3.	Suckling mouse test is used to assay eneterotoxin elaborated by *E.coli.*	( )
4.	ERV technique determines incipient spoilage in meats as well as in predicting refrigerator shelf life.	( )
5.	Bacteriological spoilage of eggs is known as 'Rotting'.	( )
6.	'Leakage type' spoilage of canned food is by the flora of spore fermenting organisms.	( )

7.	The antibiotic Nisin produced by *Staphylococcus lactis* was the first antibiotic permitted in foods to prevent spoilage of cheese.	( )
8.	Nitrate used for curing meats reacts with secondary amines to produce nitrosamines which are carcinogenic.	( )
9.	UV lights are powerful bactericidal agent, with most effective wavelength being about 3600 Angstrom.	( )
10.	Redurization extends the shelf life of sea foods and meats by 2 to 6 fold.	( )
11.	Fast freezing is done by quickly lowering temperature to about -20^{0}C within 30 minutes.	( )
12.	Slow freezing is the process by which desired temperature is achieved within 3-72 hrs.	( )
13.	Microbial count on the surface of New York dressed chicken when chilled in ice water for 45 minutes is reduced by 0.5 $\log_{10}$.	( )
14.	Impedance method is satisfactory for detection of Salmonella in meat.	( )
15.	Brine with Acidobacta preserves beef cuts better but the sensory quality is inferior.	( )
16.	Two steps mucosal competitive exclusion flora treatment to broiler chicks is not useful in reducing Salmonella contamination in finished broilers.	( )
17.	*Listeria monocytogenes* colonizes more in caeca and spleen when compared to duodenum and liver.	( )
18.	*Salmonella typhimurium* organisms get destroyed during preapartion of Kokaric.	( )
19.	Buffered lactic acid enhances decontaminating effect of Salmonella and enhances the shelf life of chicken cut up parts, when treated with it.	( )
20.	*Salmonella enteridis* phage type-4 can survive in frozen and chilled meat and poultry products.	( )
21.	Meat cuts with a microbial load of 10^4/sq.cm when held at +5^{0}C would develop off odour after 2 days.	( )
22.	Ability to utilize low molecular weight soluble components present in meat by microorganisms is not different under aerobic and anaerobic conditions.	( )
23.	Salmonella organisms fail to grow in meat at pH 4.0 irrespective of the temperature and nature of the acids.	( )
24.	Certain Psychrotrophic bacteria can grow even at 0^{0}C.	( )
25.	Penetration of *Salmonella typhimurium* into shell eggs vary according to their shell quality based on specific gravity.	( )

26.	Homogenization has no effect on the population reduction of bacteria in all time temperature combinations used for ultrapasteurization of egg liquid.	( )
27.	*E.coli* serotype 0:61 is commonly encountered in milk, poultry products and goat meat.	( )
28.	*Clostridium botulinum* type-A and type-E can grow at 10^{0}C and 3.3^{0}C respectively provided the pH of the media is more than 4.5.	( )
29.	*Clostridium perfringens* can grow at 50^{0}C but the optimum temperature is 43-47^{0}C. Meat cultures containing the organism including sterile filtrates of it are toxic.	( )
30.	T.A. spoilage of canned meat is due to thermophilic anaerobes producing hydrogen sulfide.	( )
31.	Greening of sausage is due to contamination of the product with homo fermentative lactobacillus and leuconostoc group of organisms.	( )
32.	*Streptococcus lactis* growing alone in milk causes a colour change in milk.	( )
33.	Bacillus organisms are important in the spoilage of many foods held above refrigeration temperature.	( )
34.	Surface ropiness of milk is caused most often by *Alcaligenes viscolactis*.	( )
35.	Spoilage of eggs is caused more by molds than by bacteria.	( )
36.	Lactoperoxidase complex in raw milk is effective against some streptococci.	( )
37.	Surface taint in butter is caused by *Pseudomonas putrifasciens*.	( )
38.	Leakage type spoilage of canned foods is by the flora of non spore forming organisms.	( )
39.	Freshly iced fish are invariably spoiled by fungi.	( )
40.	The earliest signs of organoleptic spoilage may be noted by examining the grills for presence of off-odour.	( )
41.	ERV techniques determine incipient spoilage in meats as well as in predicting refrigerated shelf life.	( )
42.	Stringy or ropy pickle occurs in vinegar pickle sausages.	( )
43.	*Streptococcus faecalis* is cause of food poisoning.	( )
44.	The food poisoning Streptococci grow between 10 to 45^{0}C.	( )
45.	Clostridium welchii is common cause of food poisoning in man.	( )
46.	Boiling for 10 minutes destroy *Clostridium botulinum* entero toxins.	( )

47.	Smoking of meat reduces the spoiling organism in the meat.	( )
48.	Smoked meat is secured mold growth because smoke deposit prevents mold.	( )
49.	Staphylococcus food poisoning do not occur through eating sea foods because in high saline water the Staphylococci do not develop so they are not present in sea food.	( )
50.	The carcass of pig affected with Swine erysipelas is rejected for human consumption.	( )
51.	A carcass of animal affected with lump jaw is discarded for human consumption.	( )
52.	Eggs are safest for consumption since egg shell limits the entry of bacteria to eggs.	( )
53.	A carcass having generalized coccidiosis lesions is discarded for consumption.	( )
54.	A pig carcass having hydatid disease of liver is discarded for human consumption.	( )
55.	Carcass of fowl with Pseudotuberculosis is condemned for consumption.	( )
56.	Carcass of fowl showing active lesions of fowl cholera is unfit for human consumption.	( )
57.	Presence of large number of *E. coli* in foods generally indicates faecal contamination.	( )
58.	*Lactobacillus thermophilus* survives milk pasteurization temperature.	( )
59.	Bacillus organisms are not important in spoilage of foods held above refrigerator temperature.	( )
60.	Both Clostridia and Corynebacterium are spore formers important food borne bacteria.	( )
61.	Micrococci are G+ve cocci which are catalase +ve in contrast to Streptococci.	( )
62.	Streptococcus lactis is most common cause of sour milk.	( )
63.	Most molds important in foods are placed in fungi imperfecti whose sex cycle is not known.	( )
64.	G-ve endotoxins in extracts of food is detected by dye reduction tests.	( )
65.	DNase produced by *Staphylococcus aureus* in foods is heat labile.	( )
66.	DMC method for microbiological examinations of foods determines both viable and non-viable cells.	( )
67.	Immunodiffusion test was first immunological technique used to detect presence of microbial products in foods on a routine basis.	( )

68.	Aerobic decomposition of proteins, peptides, amino acids leads to putrefactions.	( )
69.	Lipids are hydrolyzed by microbial lipase to glycerol and fatty acids in food.	( )
70.	The heat resistance of organisms is expressed in term of their thermal death time.	( )
71.	Green rots of eggs is caused by Serratia.	( )
72.	Mold spoilage of egg is generally referred as 'pin spots'.	( )
73.	Leakage type spoilage of canned food is by spore forming organism.	( )
74.	*Vibrio parahaemolyticus* food poisoning is contracted almost solely from sea foods.	( )
75.	*Micrococcus radiodurans* and *Micrococcus radiophilus* are less resistant to radiations then spore formers.	( )
76.	*Yersinia eneterocolitica* is a psychrotroph and can cause food borne outbreaks.	( )
77.	In general bacteria require low values of a_w for growth than fungi.	( )
78.	G-ve bacteria require higher values of a_w than G+ve bacteria.	( )
79.	Most of spoilage bacteria grow well, if a_w values are below 0.9.	( )
80.	Steaks stored in 15%Co_2 +85% air for 20 days had lower counts than those stored in air.	( )
81.	Fish stored in80%Co_2 +20% air for 14 days at 35^{0}C had double the surface count of the air control.	( )
82.	Hot boned hams were found to contain higher APC than cold-boned ham.	( )
83.	Poultry represents the single most important source of Salmonellae to man.	( )
84.	Upon thawing food frozen by the slow freezing method tend to lose more drip than quick frozen foods held for same period of time.	( )
85.	G-ve rods are more resistant withstand the lethal effect of freezing than G+ve cocci.	( )
86.	More microorganisms are destroyed at -4^{0}C than at -15^{0}C or below.	( )
87.	PA 3679 is used as a test organism for the evaluation of thermal process time in canned meats.	( )
88.	In general, more spoilage of eggs is caused by molds than bacteria.	( )

89.	Green rots in eggs are caused by *Pseudomonas fluorescens*	( )
90.	Burnt or caramel flavour in milk is produced by *Bacillus cereus.*	( )
91.	Streptococcus lactis causes a yellow colour in the cream layer of milk.	( )
92.	Blue milk is usually caused by *Serratia marscence.*	( )
93.	Fish from northern waters contain more of mesophilic bacteria than the fish from tropical waters.	( )
94.	Because of the presence of curing salts, these meats are spoiled mainly by G+ve bacteria, yeats and molds.	( )
95.	Souring in ham is caused by salt tolerant thermophillic bacteria.	( )
96.	The vacuum packaged meats are mainly spoiled by lactic acid bacteria.	( )
97.	Fish showing highly offensive odour particularly just under the skin, often on only on one side, may still be fresh based on other criteria.	( )
98.	Fish with plate count less than 10^5/gm, may be incremented with case of food poisoning.	( )
99.	In frozen fish products Enterococci are better indicators of faecal contamination than Coliforms.	( )
100.	Clostridium type-E cannot produce toxin at a temperature below 4^0C.	( )
101.	When the total count does not exceed 250,000/gm and *E. coli* count is less than 20/200gm/fish the plant sanitation standards is considered to be good.	( )
102.	Normal potable water used in meat processing plant contains around 100 bacteria/ml.	( )
103.	Above pH 4.5, only heat resistant spore forming bacteria can become significant as spoilage agents.	( )
104.	*Clostridium botulinum* will not grow in meat at ph 4.6.	( )
105.	*Bacillus stearothermophilus* is more heat resistant than *Clostridium botulinum*	( )
106.	Halophilic bacteria may not grow at an a_w of 0.75.	( )
107.	Presence of sugar in high concentration has no protective action against bacterial spores during canning.	( )
108.	Spoilage of meat is a surface phenomenon unless comminution and mechanical tenderizing have destroyed the tissue integrity.	( )
109.	Pseudomonas group of organisms use glucose first and Alteromonas amino acids, for their growth in meat.	( )

110.	*Bronchothrix thermosphacta* occurs only in vacuum packed stored meat.	( )
111.	At pH, more than 6.0 *Alteromonas putrifaciens* produces H_2S leading to greening of meat.	( )
112.	Soy extended ground meat spoils faster due to increase in 0.3-0.4 unit's pH.	( )
113.	UV rays are used for surface sterilization due to poor penetration power.	( )
114.	Gamma rays are used for depth sterilization of food due to excellent penetration power and short wavelength.	( )
115.	Heat treatment of foods is essential before subjecting to radiation because it is essential to destroy enzyme activity.	( )
116.	Common salt is used as preservative in meat it causes shrinkage/pleomolysis or the bacterial cell and also makes water unavailable for their growth.	( )
117.	The pH in oysters decreases as the extent of spoilage increases because it contains high carbohydrate which is broken down to different acids and therefore pH falls.	( )
118.	Endospores remain viable in presence of the antibotulinal effect and can germinate when transferred to nitrite free disc.	( )
119.	Antibotulinal effect of nitrites is more pronounced in acidic pH due to overall increase in the undissociated HNo_2.	( )
120.	A good quality meat is always sterile.	( )
121.	The pH of MDM is higher than conventionally processed meat.	( )
122.	The total number of organisms in liver, kidneys, heart etc. are generally higher than respective species meats.	( )
123.	Cocci are more heat labile than rods.	( )
124.	Clumps are less heat resistant than separate organism.	( )
125.	IMF produced by adsorption are more inhibitory to microbes than by desorption.	( )
126.	1 KGY is equal to 10^5 rads.	( )
127.	Viruses are considerably less resistant to radiation than bacteria.	( )
128.	Radiation resistance or microbes increases at colder temperature and decreases at higher temperature.	( )
129.	Radurization with 0.05-0.1 M rads extends shelf life or fresh meats and poultry by about 19-20 days.	( )
130.	Approx. Eh value of minced meat is +200mv and of solid meats is +500mv.	( )
131.	Egg white contains microbial inhibitor substance called Hydrolysates.	( )

132.	Under high humidity refrigerated meats are spoiled invariably by bacteria, essential feature of which is surface and lime.	( )
133.	Chlortetracycline and Oxytetracycline are used at the level of 100 ppm to control bacterial spoilage in uncooked refrigerated poultry and are heat sensitive.	( )
134.	Staphylococcus enterotoxin in milk can be destroyed by boiling at 100^{0}C for 10 min.	( )
135.	Aflatoxin B_1 is found in milk of cow consuming the toxin.	( )
136.	In MBR test, if it takes less than 20 min. for decolourization, the plate count will be less than 20 million/ ml.	( )
137.	The standard plate count per milliliter of pasteurized milk should not exceed 30000	( )
138.	Milk with disc rating between 2 and 4 by resazurin test is considered suitable for liquid consumption.	( )
139.	Pasteurized milk is considered safe, if coliforms are absent in 1:10 dilution.	( )
140.	Raw milk is considered satisfactory, if coliforms are absent in 1:1000 diluation.	( )
141.	Formation of lactic acid and gas are changes that commonly occur in milk.	( )
142.	*Streptococcus lactis* is the most rapid producer of acid in milk.	( )
143.	Yoghurt starter is a mixed culture of *Streptococcus thermophilus* and *Lactobacillus bulgaricus* in a 1:1 ratio.	( )
144.	To prepare dry sausages from pork, *Pediococcus cerevisiae* is used as fermenting organisms.	( )
145.	Raw ingredient for kumiss is raw cow's milk.	( )
146.	For blue cheeses such as Requefort, the curd is inoculated with spores of *Rhizopus requuefortii.*	( )
147.	The maximum tolerance level of sodium nitrite in meat curing preparation is320ppm.	( )
148.	Sodium nitrite is used in curing formulae for meats as they stabilize red meat colour, inhibit some spoilage and contribute to flavour development.	( )
149.	Nisin is the most widely used antibiotic for food preservation but it is not permitted in foods in the USA and Canada.	( )
150.	D-value is the decimal reduction time or the time required to destroy 50% of the organism.	( )
151.	Dry milk products should not contain at any time a standard plate count over 30000 per gram or coliform over 90 per gram.	( )

152.	The maximum permissible limit of Salmonella in frozen meat is $2x10^2$ organism per gram.	( )
153.	As many as 1000 *E.coli* are permissible in frozen fish.	( )
154.	*Clostridium nigrificans* is a sulfide spoiler.	( )
155.	*Bacillus coagulans* belongs to thermophilic flat sour group.	( )
156.	Non spore formers imparts acidic odour to spoiled canned foods.	( )
157.	Thermophilic anaerobes imparts cheesy or butyric odour to spoiled canned foods.	( )
158.	Antimicrobial defense afforded by shell membrane is partly physical and partly chemical.	( )
159.	Usually organisms on egg shell do not multiply because of the low level of available water.	( )
160.	160. Micro-organisms that have gained access to the inside of the shell can be killed by holding eggs at low temperature (0^0C).	( )
161.	Salmonella are more heat resistant in egg yolk as compared to egg white.	( )
162.	The main purpose of pasteurizing egg products is to create a sterile product.	( )
163.	The first step in setting criteria involves compilation of a list of hazardous micro-organisms.	( )
164.	On surface of sanitized meat service equipment the residual bacterial count should not exceed 500.	( )
165.	International Food Standards are evolved by joint FAO/WHO/ Food standards commission.	( )
166.	Microbiological specifications is the minimum acceptable number of micro-organisms in a food.	( )
167.	Microbiological standards is that part of law which designates the maximum acceptable number of microorganisms.	( )
168.	The finalized International standards are published in a Food Codex Alimetarius.	( )
169.	International Commision on Microbiological Specifications for foods (ICMSF) is a non-voluntary body which establishes sampling plans and methods of analysis.	( )
170.	Plate counts on food stuff are approximate only.	( )
171.	While choosing standards there should be no definite relations between standards and hazards.	( )
172.	Tolerances are never included in the standards to account for in accuracies of sampling and analysis.	( )

173.	Micro-organisms grow best at pH values around 7.0.	( )
174.	The a_w of most fresh food is below 0.99.	( )
175.	Molds are filamentous fungi that grow in the form of a tangled mass.	( )
176.	Typical yeast cells range from 5 to 8 μm in diameter.	( )
177.	Whole poultry tends to have a higher microbial count than cut up poultry.	( )
178.	ERV is an important value for determining incipient spoilage in meat.	( )
179.	Presence of Coliforms in food is an indicator of faecal contamination.	( )
180.	The concept of HACCP came from ICMSF.	( )
181.	Final ultimate pH of the poultry meat is about 5.6 and above.	( )
182.	The primary nitrogen sources utilized by heterotrophic microorganisms are amino acids.	( )
183.	Eggs contain lysozyme and this enzyme, along with conalbumin, provides fresh eggs with a fairly efficient antimicrobial system.	( )
184.	Ovotransferrin appears to be the inhibitory substance in raw egg white that inhibits *Salmonella enteritidis.*	( )
185.	In log phase of bacterial growth curve bacterial cells increase in size.	( )
186.	The lowest temperature at which a microorganism has been reported to grow is -34°C	( )
187.	Members of the genera Paenibacillus, Bacillus, and Clostridium, are found in meats of all types.	( )
188.	188. Lysozyme of egg albumin causes lysis of mucopeptide rich cellwall of G-ve organisms.	( )
189.	Green rot in eggs is caused by Serratia spp.	( )
190.	In general, Arcobacter spp. appear to be more common among poultry than red meat products.	( )

Q.4 Match the following

A.

A		B		
1.	TA spoilage of canned	a.	*Desulfotomsculum nigrifaciens*	( )
2.	Sulphur stinker spoilage of canned food	b.	*Clostridium thermosacchrolyticum*	( )

3.	Burnt of caramel flavour of milk	c.	*Pseudomonas putrefaciens*	( )
4.	Surface taint of butter	d.	*Serratia* spp	( )
5.	Green rots of eggs	e.	*Aspergillus repens* and *Pencillium*	( )
6.	Red rots of eggs	f.	Thermoduric and spore formers	( )
7.	Phosphorescence of meats	g.	*Streptcoccus lactis* var. multigenes	( )
8.	Buttons (mold colonies growing on milk surfaces)	h.	Micrococci	( )
9.	Thickening of sweetened condensed milk	i.	*Photobacterium* spp	( )
10.	Bacteria that survive pasteurisation	j.	*Pseudomonas fluorescence*	( )

B.

A		B		
1.	Yeast	a.	Minimum pH 4.4 for growth	( )
2.	*E. coli*	b.	5.9-6.2	( )
3.	Thiobacillus thioxidans	c.	Minimum pH 2.5 for growth	( )
4.	Molds	d.	Minimum pH 4.5 for growth	( )
5.	Salmonella	e.	Thamnidium and Mucor	( )
6.	Minimum microbial quality of oysters	f.	Minimum pH 1.0 for growth	( )
7.	Whiskers in beef	g.	Sporotrichum	( )
8.	Black spot in beef	h.	Penecillium	( )
9.	Green patches	i.	Minimum pH 1.5-2.0 for growth	( )
10.	White spots	j.	Cladosporium	( )

C.

A		B		
1.	Greening in fresh red meat	a.	CFU	( )
2.	Black spot in cured meat	b.	<0.5 million/g	( )
3.	Statistical determination of viable cells	c.	MPN	( )
4.	Microbial population in the food	d.	*Pseudomonas mephitica*	( )
5.	Standards TVC in fresh meat	e.	*Clostridium nigrificans*	( )

6.	Standard TVC in raw meat products	f.	Halophilic	( )
7.	Food-poisoning bacteria	g.	200-10,000/g	( )
8.	Salt loving bacteria	h.	Staphylococcus aureus	( )
9.	Dry loving	i.	Role of microorganisms in the spoilage	( )
10.	L. Pasteur	j.	Xerophilic	( )

D.

A		B		
1.	Coliforms	a.	Enterotoxigenic	( )
2.	Enterobacteriaceae	b.	Ground beef	( )
3.	Enterococci	c.	Meat preservation	( )
4.	Staphylococcus aureus	d.	Faecal contamination	( )
5.	E.coli	e.	Food Safety Management System	( )
6.	Clostridium botulinum	f.	Defective heat/processing	( )
7.	Campylobacter jejuni	g.	Rare cause of disease in human	( )
8.	Trichinella spiralis	h.	Sanitation	( )
9.	HACCP	i.	Poultry meat	( )
10.	Hurdle Technology	j.	Raw pork	( )

Food Microbiology Key

Q.No. 1	
1.d	38.d
2.d	39.d
3.d	40.d
4.d	41.d
5.c	42.c
6.d	43.a
7.d	44.b
8.c	45.c
9.d	46.a
10.c	47.c
11.b	48.c
12.a	49.d
13.c	50.d
14.b	51.c
15.b	52.c
16.b	53.d
17.a	54.d
18.a	55.c
19.a	56.d
20.c	57.d
21.b	58.c
22.b	59.d
23.a	60.d
24.c	61.d
25.a	62.a
26.d	63.a
27.d	64.c
28.a	65.a
29.c	
30.d	
31.d	
32.a	
33.c	
34.c	
35.a	
36.c	
37.c	

Q.No. 2			
1.	*Pseudomonas fluorescence*	37.	Lysozyme
2.	Red rots	38.	15-25°C
3.	37°C	39.	Exoenzymes
4.	65°C	40.	32-45°C
5.	Flipper	41.	Putrifactive changes
6.	Springer	42.	Hydrogen sulphide
7.	-20°C within 30 minutes	43.	Sugar
8.	Thermal death time	44.	*Salmonella typhimurium*
9.	D-value	45.	1-6 hrs
10.	*Satphylococcus aureus*	46.	2-48 hrs
11.	Bacteria	47.	Nitrogenous compounds
12.	Fungi	48.	Sheep
13.	Halophiles	49.	Pseudomonas
14.	Halodurics	50.	Pseudomonas and Achromobacter
15.	6.7°C	51.	Gayon
16.	10°C	52.	Lister
17.	Spoilage	53.	Miquel
18.	Bacteriocin	54.	Von Geuns
19.	Antagonistic products	55.	Sehmidt-Nielsen
20.	0.99	56.	B.W. Hammer
21.	7.0	57.	Donk in 1917
22.	0.93	58.	1888
23.	0.86	59.	Van Ermengen
24.	0.75	60.	Aspergillus flavus in 1960
25.	Low temperature and catalase negative	61.	Water activity (aw)
26.	Sugar	62.	Radurization
27.	*Clostridium botulinum*	63.	Microaerophiles
28.	Lower heat resistance of the spores of the latter	64.	Osmophiles
29.	*Streptococcus lactis* var maltigenes	65.	Thermophiles
30.	10^7/gm	66.	Lactoperoxidase
31.	10^8/gm	67.	Anticoliform factor
32.	100,000/gm	68.	Lysozyme
33.	500,000/gm	69.	Cinnamon acid
34.	Minimum	70.	Eugenol
35.	Short high temperature	71.	Capsules or slime
36.	4.5	72.	Bacillus and Clostridia

73.	Aspergillus molds	110.	Acetobacter
74.	Anaerobic	111.	Photobacterium
75.	Aerobic	112.	*Yersinia enterocolitica*
76.	Gill	113.	Twice/more
77.	Scales	112.	*Yersinia enterocolitica*
78.	F. Appert	113.	Twice/more
79.	0.97	112.	*Yersinia enterocolitica*
80.	0.70	113.	Twice/more
81.	0.75	114.	Osmophilic
82.	0.96	115.	Association of Official Analytical Chemists
83.	Above 6.0	116.	*Escherichia coli (Biotype-I)*
84.	0.99	117.	Thermophiles
85.	Psychrotrophs and Mesophils	118.	Aerobes
86.	-10 to -18^{0}C	119.	Microscope using oil immersion objective
87.	Pseudomonas	120.	Leistner and Rodel
88.	Lactobacillus	121.	Lower
89.	Lactobacilli	122.	Enzyme Linked Immuno Sorbent Assay
90.	Eneterobacter	123.	Schmidt-Nielsen
91.	Decorboxylating	124.	Selective media
92.	-18^{0}F	125.	Salmonella
93.	62.8^{0}C	121.	Lower
94.	Poultry	122.	Enzyme Linked Immuno Sorbent Assay
95.	-0.5^{0}C to 4^{0}C	123.	Schmidt-Nielsen
96.	-8^{0}C	124.	Selective media
97.	Excites and ionizes	125.	Salmonella
98.	*Micrococcus radiodurans*		
99.	15-50		
100.	120		
101.	*Streptococcus lactis*		
102.	*Aspergillus repens*		
103.	*Leuconostoc mesenteroides*		
104.	4.6		
105.	*C. perfringens* type-c		
106.	Self stable canned cured meats		
107.	*Staphylococcus aureus*		
108.	Pseudomonas		
109.	Salmonella		

Q. No. 3									
1.	F	41.	T	81.	F	121.	T	161.	T
2.	F	42.	T	82.	T	122.	F	162.	F
3.	T	43.	T	83.	T	123.	F	163.	T
4.	T	44.	T	84.	T	124.	F	164.	T
5.	T	45.	T	85.	F	125.	T	165.	T
6.	F	46.	T	86.	T	126.	T	166.	F
7.	F	47.	T	87.	T	127.	F	167.	T
8.	T	48.	F	88.	F	128.	T	168.	T
9.	F	49.	F	89.	T	129.	T	169.	F
10.	T	50.	F	90.	F	130.	F	170.	T
11.	T	51.	F	91.	F	131.	F	171.	F
12.	T	52.	F	92.	F	132.	T	172.	F
13.	T	53.	T	93.	F	133.	F	173.	T
14.	T	54.	T	94.	T	134.	F	174.	F
15.	F	55.	T	95.	F	135.	F	175.	T
16.	F	56.	T	96.	T	136.	F	176.	T
17.	T	57.	T	97.	T	137.	T	177.	F
18.	T	58.	T	98.	F	138.	F	178.	T
19.	T	59.	F	99.	T	139.	T	179.	T
20.	T	60.	F	100.	F	140.	F	180.	T
21.	T	61.	T	101.	T	141.	T	181.	T
22.	T	62.	T	102.	T	142.	T	182.	T
23.	F	63.	T	103.	F	143.	T	183.	T
24.	T	64.	F	104.	F	144.	T	184.	T
25.	T	65.	F	105.	T	145.	F	185.	F
26.	F	66.	T	106.	F	146.	F	186.	T
27.	F	67.	T	107.	F	147.	F	187.	T
28.	T	68.	F	108.	T	148.	T	188.	T
29.	F	69.	T	109.	T	149.	T	189.	F
30.	F	70.	T	110.	F	150.	F	190.	T
31.	F	71.	F	111.	T	151.	T		
32.	F	72.	T	112.	T	152.	F		
33.	T	73.	F	113.	T	153.	F		
34.	T	74.	T	114.	T	154.	T		
35.	F	75.	F	115.	T	155.	T		
36.	T	76.	T	116.	T	156.	T		
37.	T	77.	F	117.	T	157.	T		
38.	T	78.	T	118.	T	158.	T		
39.	F	79.	F	119.	T	159.	T		
40.	T	80.	T	120.	F	160.	F		

Q.No. 4							
A		B		C		D	
1.	b	1.	c	1.	d	1	d
2.	a	2.	a	2.	e	2.	f
3.	g	3.	f	3.	c	3.	h
4.	c	4.	i	4.	a	4.	a
5.	j	5.	d	5.	b	5.	b
6.	d	6.	b	6.	g	6.	g
7.	i	7.	e	7.	h	7.	i
8.	e	8.	j	8.	f	8.	j
9.	h	9.	h	9.	j	9.	e
10.	f	10.	g	10.	i	10.	c

7

Poultry & Egg Products Technology

Q.1 Multiple type questions

1. Ideal temperature time combination for scalding broilers is
   a. 53°C for 120 seconds
   b. 60°C for 60 seconds
   c. 93°C for 5 seconds
   d. 60°C for 120 seconds
2. What will be your Post-Mortem decision in case of 'Ascitis' in poultry
   a. Total unfit for human consumption
   b. Fit for human consumption
   c. Partially condemned and passed for human consumption
   d. None of the above
3. Post mortem judgment for poultry carcass showing lesions of fowl typhoid is
   a. Total rejection
   b. Conditionally passed
   c. decision as per carcass state
   d. Rejection of local parts
4. Thermostabilization of eggs is done at:
   a. 550C for 15 minutes
   b. 580C for 10 minutes
   c. Both of the above
   d. None of the above
5. Egg is a poor source of
   a. Vitamin-C
   b. Vitamin-D
   c. Vitamin-A
   d. All of the above
6. Red rot of eggs is caused by
   a. *Proteus melenovogenes*
   b. *Pseudomonas maltophila*
   c. *Pseudomonas flourescens*
   d. *Serratia marcescens*
7. Hay odour in eggs is caused by
   a. *Achromobactor porolens*
   b. *Pseudomonas mucidolens*
   c. *Pseudomonas graveolens*
   d. *Enterobacter cloacae*

8. Calcium carbonate content of egg shell is about
   a. 94%
   b. 4%
   c. 1%
   d. 50%
9. Carbohydrate content of poultry meat is
   a. 5-6%
   b. 9-10%
   c. 15-16%
   d. 1-2%
10. Duration of fasting before slaughtering poultry should not exceeds
    a. 12 hours
    b. 23 hours
    c. 36 hours
    d. 48 hours
11. Singeing is generally performed over
    a. Blue flame
    b. White flame
    c. Grey flame
    d. None of the above
12. The absorption of meat fat is
    a. 80%
    b. 90%
    c. 100%
    d. 75%
13. Castrated male chicken is called
    a. Capon
    b. Pullet
    c. Cockrale
    d. Clutch
14. Most of the fat in turkey is located at
    a. Abdomen
    b. Subcutaneously
    c. Intramuscularly
    d. Intermuscular
15. Which among is the red meat bird
    a. Goose
    b. Duck
    c. Turkey
    d. Guinea fowl
16. Daily need of energy for an Indian individual is
    a. 1500-2000 Calorie
    b. 3500-4000 Calorie
    c. 2500-3000 Calorie
    d. None of the above
17. Which is the incorrect answer regarding thermostabilization of eggs
    a. 49 °C for 35 min.
    b. 54 °C for 15 min.
    c. 56 °C for 10 min
    d. 60 °C for 1 min.

18. Indicate the Nodal Agency for developing a strong and vibrant food processing sector in India

a. APEDA
b. MPEDA
c. MFPO
d. BIS

19. MPEDA functions under the Ministry of

a. Rural Development
b. Food processing Industry
c. Commerce
d. Consumer affair

20. Poultry Birds should be kept off feed for-------- but sufficient water should be provided before slaughter

a. 12 hrs
b. 24 hrs
c. 36 hrs
d. No need to put on off feed

21. Poultry with morbid condition due to clinical evidence of a contagious disease, heat stroke or traumatic injury in ante-mortem examination will be declared as

a. Fit for slaughter
b. Unfit for slaughter
c. Suspects
d. Delayed slaughter

22. Core temperature in patties preparation must reach to

a. 72°C
b. 64°C
c. 60°C
d. 80°C

23. Scalding temperature for broilers is

a. 55°C for 1.5 min.
b. 60°C for 10 min.
c. 63°C for 1 min
d. 60°C for 2 min.

24. Eggs ranks ----------- as a natural source of Vit-D

a. Third
b. Second
c. First
d. Fourth

25. Blood generally amounts .... of the total carcass weight in poultry.

a. 9 to 10%
b. 5.5 to 6.6%
c. 3.3 to 4.8%
d. None of the above

26. Irradiation enhances the shelf life of poultry meat by

a. DNA fragmentation
b. Cell wall lysis
c. Protein destruction
d. All of the above

27. In baked goods, eggs act primarily as

    a. Leavening agent
    b. Emulsifying agent
    c. Clarifying agent
    d. None of the above

28. The rate of loss of the following is very high in the eggs during the first few hours of production

    a. Oxygen
    b. Carbon dioxide
    c. Nitrogen
    d. None of the above

29. Avidin, a toxic protein , can combine with the following essential vitamin and render it unavailable

    a. Vitamin-A
    b. Biotin
    c. Vitamin-C
    d. None of the above

30. The yolk begins to coagulate at

    a. 50°C
    b. 60°C
    c. 65°C
    d. 70°C

31. The cholesterol level of the egg yolk

    a. Increased with an increase in the degree of unsaturation of the dietary fat of laying hens
    b. Increased with an decrease in the degree of unsaturation of the dietary fat of laying hens
    c. Is not influenced by the quality of dietary fat of the laying hens
    d. None of the above

32. Preservation of meat by ionizing radiation is known as

    a. Chemical sterilization
    b. Cold sterilization
    c. Warm sterilization
    d. None of the above

33. Egg wash water temperature for warming the dirt and stains from shell surface shall be

    a. At least 60°C for adequate cleaning without development of thermal cracks
    b. At least 35°C for adequate cleaning without development of thermal cracks
    c. At least 65°C for adequate cleaning without development of thermal cracks
    d. None of the above

34. On an average, the poultry meat contains
    a. More percent proteins than beef or lamb or pork
    b. Less percent proteins than beef or lamb or pork
    c. Equal percent proteins than beef or lamb or pork
    d. None of the above

35. As the egg ages, the weight of the yolk is increased due to
    a. Migration of water from yolk to albumen
    b. Migration of air from the air cell to yolk
    c. Migration of water from the albumen to yolk
    d. None of the above

36. Fifty percent chance of moving through normal market handling without breaking of eggs were estimated to have at least
    a. A shell thickness of 0.25 mm
    b. A shell thickness of 0.33mm
    c. A shell thickness of 0.19 mm
    d. None of the above

37. Yolk constitutes around
    a. 43-45% of whole egg
    b. 55-57% of whole egg
    c. 22-24% of whole egg
    d. 30-32% of whole egg

38. A hen's egg contains around
    a. 10-11g of protein
    b. 12-13g of protein
    c. 6-7g of protein
    d. 14-15g of protein

39. A component of egg white having antibacterial activity is known as
    a. Avidin
    b. Conalbumen
    c. Albumen
    d. Lysozyme

40. Minimum requirements for pasteurization of whole egg blends (less than 2% added non-egg ingredient) are
    a. 142°F for 3.5 min
    b. 134°F for 3.5 min
    c. 146°F for 3.5 min
    d. 132°F for 3.5 min

41. Most tender meat can be obtained from
    a. Turkey
    b. Geese
    c. Broiler chickens
    d. Ducks

42. Compared to red meats poultry meat contains a higher proportion of
    a. Saturated fatty acid
    b. Unsaturated fatty acids
    c. Both of the above
    d. None of the above
43. Freeze dried poultry products are generally stored at
    a. Ambient temperature
    b. Refrigerated temperature
    c. Freezing temperature
    d. None of the above
44. By irradiation oxidative deterioration of poultry meat is
    a. Retarded
    b. Hastened
    c. Stable
    d. None of the above
45. Chicken pickle and tandoori chicken are foods of
    a. Indigenous origin
    b. Exotic origin
    c. Synthetic origin
    d. None of the above
46. Warner Bratzler shear press is used to measure
    a. Flavour
    b. Colour
    c. Tenderness
    d. None of the above
47. Egg white proteins largely responsible for rapid whipping properties is
    a. Ovoglobulin
    b. Ovomucin
    c. Ovalbumen
    d. None of the above
48. The most commonly whipping aid added to egg white products are
    a. Sodium laurylsulfate
    b. Triethyl citrate
    c. Triacetin
    d. All of the above
49. When 390 quintals of whole egg liquid are dried approximately the following quantity of dried product is available
    a. 10 quintals
    b. 25 quintals
    c. 50 quintals
    d. 100 quintals
50. The naturally occurring pigments in egg yolk are mainly
    a. Xanthophylls
    b. Luten
    c. Zeaxanphin
    d. All of the above

51. In maillard reaction glucose combines with the following component of egg and make the product insoluble

 a. Protein
 b. Amino acid
 c. Cephalin
 d. All of the above

52. Egg contents start coagulating at

 a. 62°C
 b. 62.4°C
 c. 62.8°C
 d. 63.2°C

53. Geese are scalded at

 a. 145-149°F
 b. 150-154°F
 c. 155-159°F
 d. 160-164°F

54. Dressing percentage of geese slaughtered at 17 weeks age is about

 a. 68%
 b. 70%
 c. 72%
 d. 74%

55. For packaging broiler the approximate film sheet size for wrapping varies from

 a. 16x18 to 16x28
 b. 20x22 to 22x24
 c. 24x26 to 26x28
 d. None of the above

56. According to USDA regulation, carcasses must be chilled to atleast _____ internal temeperature

 a. 40°F
 b. 35°F
 c. 45°F
 d. 50°F

57. Coating of poultry carcasses with irish moss extract containing six percent sodium chloride approximately increases the storage life

 a. 2 fold
 b. 3fold
 c. 4 fold
 d. 5 fold

58. Formaldehyde in smoke is

 a. Carcinogenic
 b. Bacteriostatic
 c. Bactericidal
 d. None of the above

59. After smoking the cooking of poultry carcasses is made until the internal temperature of meat reaches

 a. 140°F
 b. 155°F
 c. 165°F
 d. 178°F

60. During curing carcasses should be kept around

a. 35-40°F b. 40-45°F
c. 48-50°F d. 60-65°F

61. After removing from cure solution the carcasses should be soaked in cold water for

a. 2-4 hrs b. 4-6 hrs
c. 6-8 hrs d. 8-10hrs

62. Low smoke house temperature

a. Increase the shrinkage b. Decrease the shrinkage
c. No effect on shrinkage d. None of the above

63. Maximum storage time at 0°F for smoked meat is

a. 1-2 months b. 2-4 months
c. 4-6 months d. 8-10 months

64. The part of carcass not included in turkey liver qurst is

a. Liver b. Gizzard
c. Neck d. Heart

65. In steam vac closure canning the broth or brine temperature is maintained at

a. 180°F b. 166°F
c. 170°F d. 190°F

66. The birds generally preferred for canning

a. Mature b. Young
c. Both of the above d. None of the above

67. For poultry canning lacquering of cans are essential with

a. Acid resistant b. Sulpher resistant
c. Both of the above d. None of the above

68. Flash 18 process is conducting using

a. 16 lb pressure b. 18 lb pressure
c. 20 lb pressure d. 25 lb pressure

69. Off feed prior to slaughter does not affects significantly the nutritive value of

a. Broiler breast meat
b. Broiler drumstick
c. Both of the above
d. None of the above

70. The bleeding percentage is comparatively lower in

a. Males than females
b. Females than male
c. Both of the above
d. None of the above

71. Normally the birds are bled for

a. 90-100 seconds
b. 60-90 seconds
c. 100-120 seconds
d. None of the above

72. Dry pickling feather release is achieved by piercing

a. Pons area of brain
b. Medulla area of brain
c. Third ventricles of brain
d. None of the above

73. Hot deboning of turkey reduces the cost of deboning up to the extent of

a. 40%
b. 45%
c. 50%
d. 55%

74. Because of relatively high _____ level in some MDPM, this product is not allowed to be used in infant feeding

a. Chloride
b. Fluoride
c. Iodide
d. None of the above

75. For best quality of emulsion stability the meat used is

a. Hand deboning
b. Mechanically deboned
c. Mixture of both
d. None of the above

76. The PER of the protein in MDPM

a. 1.78
b. 2.8
c. 2.28
d. None of the above

77. The dry sausage Alpine is considered as

a. Italian brand
b. Mexican brand
c. French brand
d. None of the above

78. Broiler dueling is a young duck usually under ___________weeks

a. 6 weeks
b. 8 weeks
c. 10 weeks
d. 12 weeks

79. During the curing of poultry products which of the following chemical accelerate the nitrite curing reaction

a. Sodium erythorbate
b. Sodium metabisulfate
c. Sodium ascorbate
d. None of the above

80. One product that utilizes only thigh meat either of the chicken or turkey is

a. Hamburger
b. Pashrani
c. Bologna
d. None of the above

81. The resultant greenish black layer on the yolk surface in over boiled egg is due to

a. Ferric chloride
b. Cupric sulphide
c. Ferrous sulphide
d. Ferric sulphide

82. Pasteurization by irradiation method __________the viscosity of the egg albumen

a. Increases
b. Decreases
c. No effect
d. None of the above

83. The ______ iodine value of poultry meat indicates a higher proportion of unsaturated fatty acids than the fats from red meats.

a. High
b. Low
c. Medium
d. None of the above

84. Poultry meat contains ____________proteins as compared to red meat.

a. Less
b. More
c. Equal
d. None of the above

85. Broilers meat contain

a. 302 calories/100 g
b. 200 calories/100 g
c. 151 calories/100 g
d. None of the above

86. Carcasses from young birds have ___________proportion of moisture to tissue than old ones.

a. Higher
b. Lower
c. Equal
d. None of the above

87. The slaughter hall floors must have a fall of not less than ______inches in every ten feet

a. One
b. Two
c. Three
d. None of the above

88. ANTU in rodent baits in concentration of __________% gives most effective results

a. 1.0
b. 1.5
c. 2.0
d. None of the above

89. Cathod rays are the streams of ______________________charged electrons

a. Positively
b. Negatively
c. Neutrally
d. None of the above

90. Carbondioxide _______________increases the rate of bleeding of chickens.

a. Increases
b. Decreases
c. Stable
d. None of the above

91. Stunning of poultry by electrical method is done by using _________V current for 5-10 sec.

a. 40-50
b. 50-60
c. 60-70
d. None of the above

92. Stewing is a ______________procedure of cooking meat.

a. Moist heat
b. Dry heat
c. Combination of moist and dry heat
d. None of the above

93. Poultry processing plant management where solids and waste matter are kept minimum, the waste water of 1000 birds is roughly equal to

a. The domestic sewage from 150 people
b. The domestic sewage from 25 people
c. The domestic sewage from 50 people
d. None of the above

94. When poultry carcasses are immersed in chilling solution, the rate of absorption of moisture in carcasses

a. Fastest during first few hours
b. Slowest during first few hours
c. Nil
d. None of the above

95. The rate of bleeding of chicken is

a. Slow down by CO_2 stunning
b. Speed up by CO_2 stunning
c. Not influenced by CO_2 stunning
d. None of the above

96. The most widely followed method of slaughter in modern poultry processing operation is

a. Decapitation of head
b. Piercing through the brain
c. Modified kosher method
d. None of the above

97. Semi or soft or slack scalding is done by immersing poultry carcasses in water at

a. 50.5-54.4°C for 90-120 seconds
b. 71-82°C for 90-120 seconds
c. 58.5-60°C for 90-120 seconds
d. None of the above

98. During canning if mechanical vacuum closure is employed, is suggested to have

a. Minimum vacuum of 12 inches and maximum machine vacuum 15 inches
b. Minimum vacuum of 20 inches and maximum machine vacuum 8 inches
c. Minimum vacuum of 5 inches and maximum machine vacuum 8 inches
d. None of the above

99. Counter flow tumble chillers are essentially the same as the parallel flow system except that

a. The water remains still without circulating
b. The water circulates in the same direction of the carcasses
c. The water circulates in opposite direction of the carcasses
d. None of the above

100. Recommended level of concentration of chlorine as a sanitizer for use in food processing plants is

a. 100 ppm
b. 200ppm
c. 300ppm
d. 400ppm

101. Retort processing of whole chicken in canning, the cans water cooled to a temperature

a. Below 40°F so as to prevent the rusting of cans
b. Below 10°F so as to prevent the rusting of cans
c. Below 95°F so as to prevent the rusting of cans
d. None of the above

102. Green rot in egg is caused by

a. *Acinetobacter* spp.
b. *Pseudomonas fluorescens*
c. *Proteus melanovogenes*
d. *Serratia* spp

103. Colourless rot in egg is caused by

a. *Acinetobacter* spp.
b. *Pseudomonas fluorescens*
c. *Proteus melanovogenes*
d. *Serratia* spp

104. Black rot in egg is caused by

a. *Acinetobacter* spp.
b. *Proteus melanovogenes*
c. *Pseudomonas fluorescens*
d. *Serratia* spp

105. Mustiness in eggs is caused by bacteria:

a. *Achromobacter perolens*.
b. *Pseudomonas mucidolens*
c. *Pseudomonas graveolens*
d. All of the above

106. Fishy odour in eggs is caused by

a. *Enterobacter cloacae*
b. *Pseudomonas mucidolens*
c. *Achromobacter perolens*
d. *Escherichia coli*

107. Fungal black rot in eggs is caused by

a. Cladosporium
b. Penicillium
c. Sporotrichum
d. Mucor

108. Fungal dark green spots in eggs is caused by

a. Penicillium
b. Cladosporium
c. Sporotrichum
d. Mucor

109. Steam generators are recommended to pasteurize

a. Egg rolls
b. Egg white
c. Whole egg liquid
d. All of the above

110. The high hydrostatic pressure technology of egg processing generally use

a. 100-1000 MPa pressure
b. 1000-2000 MPa pressure
c. 2000-2500 MPa pressure
d. None of the above

111. The gamma radiation sources commonly used on commercial egg plants are

a. Cobalt 60
b. Cesium 137
c. Both of the above
d. None of the above

112. Freeze dried scrambled eggs can be categorized into

a. Refrigerated egg products
b. Frozen egg products
c. Dehydrated egg products
d. Specialty egg products

113. As per egg product inspection act, Pasteurization of all egg products is

a. Mandatory
b. Voluntary
c. No provision
d. None of the above

114. Eggs can be used as thickening agent in many food formulations like

a. Cakes
b. Pie fillings
c. Cream puddings
d. All of the above

115. Eggs are used as binding agent in

a. Cutlets
b. Chops
c. Both of the above
d. None of the above

116. Name the proteins present in egg acts as an excellent emulsifying agent

a. Avidin
b. Lecithin
c. Vitelline
d. None of the above

117. Eggs are extensively used as leavening agent in baked products such as

a. Cakes
b. Muffins
c. Both of the above
d. None of the above

118. Egg and egg products are ______________ foaming agents

a. Good
b. Bad
c. Cannot be utilized as
d. None of the above

119. The pH of egg increases during storage due to

a. Oxygen escape
b. Carbon dioxide escape
c. Both of the above
d. None of the above

120. In good quality egg air cell size should be

a. 2-3cm
b. 3-4 cm
c. 5-6 cm
d. 6-7 cm

121. Lime treated eggs can be stored for ________weeks at room temperature

a. 1-2
b. 2-3
c. 3-4
d. None of the above

122. Oiling preserves the eggs by CO_2 and other changes
   a. Preventing evaporation of carbon dioxide
   b. Preventing evaporation of water
   c. Both of the above
   d. None of the above

123. Free flowing agent used in dried whole egg or yolk solids is
   a. Sodium metasilicate
   b. Sodium citrate
   c. Sodium thiosulphate
   d. Sodium silicoaluminate

124. Name the speciality egg product
   a. Sugared yolk
   b. Whole eggs
   c. Egg crepes
   d. Salted whole eggs

125. On freezing, egg yolk becomes gummy and thick due to
   a. Gelation
   b. Desugarization
   c. Foaming
   d. Emulsification

126. Browning can be prevented by removing glucose through fermentation by yeast or with commercial enzymes and the process is called
   a. Gelation
   b. Desugarization
   c. Foaming
   d. Emulsification

127. Eggs can be used as thickening agent in many food formulations like
   a. Sugared yolk
   b. Cakes
   c. Whole eggs
   d. Salted whole eggs

128. In mayonnaise egg yolk acts as an
   a. Emulsifier
   b. Thickner
   c. Whitner
   d. Binder

129. Hough Unit for good quality egg
   a. 72
   b. 60
   c. 65
   d. 70

130. 10% solution of sodium silicate is commonly called
   a. Water glass
   b. Soda water
   c. Pinch water
   d. Lime water

131. How long the oiled eggs can be preserved at room temperature.

a. 4weeks
b. 3 weeks
c. 1 week
d. 2 weeks

132. An average weight of hen egg is about

a. 50g
b. 52g
c. 57g
d. 70g

133. Egg shell is mainly composed of

a. Calcium bicarbonate
b. Calcium silicate
c. Calcium sulphate
d. Calcium carbonate

134. Location of air cell in whole egg is

a. Large end
b. Periphery
c. Narrow end
d. Middle

135. Water test of egg is mainly based on

a. Density
b. Width
c. Weight
d. Height

136. Most commonly used test to determine the spoilage of egg is

a. Weight test
b. Colour test
c. Candling test
d. Water test

137. Vitellin is the protein present in

a. Egg yolk
b. Egg White
c. Egg Challazae
d. Egg Shell

138. Egg is the rich source of all known vitamins except

a. Vitamin D
b. Vitamin B
c. Vitamin A
d. Vitamin C

139. Shape index of good quality egg ranges from

a. 70-75
b. 60-65
c. 40-45
d. 50-55

140. The contribution of egg albumin in whole egg is about

a. 65 %
b. 45%
c. 35%
d. 55%

141. The primary phospholipid present in egg is

a. Cholestrol
b. Avidin
c. Phosphvitin
d. Lecithin

142. The absorption of meat fat is

a. 80%
b. 90%
c. 100%
d. 75%

143. Castrated male chicken is called

a. Capon
b. Pullet
c. Cockrale
d. Clutch

144. Most of the fat in turkey is located at

a. Abdomen
b. Subcutaneously
c. Intramuscularly
d. Intermuscular

145. Which among is the red meat bird

a. Goose
b. Duck
c. Turkey
d. Guinea fowl

146. Daily need of energy for an Indian individual is

a. 1500-2000 Calorie
b. 3500-4000 Calorie
c. 2500-3000 Calorie
d. None of the above

147. Which is the incorrect answer regarding thermostabilization of eggs

a. 49°C for 35 min.
b. 54°C for 15 min.
c. 56°C for 10 min
d. 60°C for 1 min.

148. Indicate the Nodal Agency for developing a strong and vibrant food processing sector in India

a. APEDA
b. MPEDA
c. MFPO
d. BIS

149. MPEDA functions under the Ministry of

a. Rural Development
b. Food Processing Industry.
c. Commerce
d. Cunsumer affair

150. Poultry Birds should be kept off feed for-------- but sufficient water should be provided before slaughter

a. 12 hrs
b. 24 hrs
c. 36 hrs
d. No need to put on off feed

151. Poultry with morbid condition due to clinical evidence of a contagious disease, heat stroke or traumatic injury in ante-mortem examination will be declared as

a. Fit for slaughter
b. Unfit for slaughter
c. Suspects
d. Delayed slaughter

152. Core temperature in patties preparation must reach to

a. 72°C
b. 64°C
c. 60°C
d. 80°C

153. Scalding temperature for broilers is

a. 55°C for 1.5 min.
b. 60°C for 10 min.
c. 63°C for 1 min
d. 60°C for 2 min.

154. Eggs ranks ---------- as a natural source of Vit-D

a. Third
b. Second
c. First
d. Fourth

155. Blood generally amounts .... of the total carcass weight in poultry

a. 9 to 10%
b. 5.5 to 6.6%
c. 3.3 to 4.8%
d. None of the above

156. Irradiation enhances the shelf life of poultry meat by

a. DNA fragmentation
b. Cell wall lysis
c. Protein destruction
d. All of the above

157. Egg shell matrix is made up of ____________ protein

a. Myosin
b. Collagen
c. Nebulin
d. All of the above

158. 'Pin cushions' in egg are produced by

a. Molds
b. Yeast
c. Chlamydia
d. All of the above

159. Temperature less than 10°C may cause ________________ of eggs

a. Heating
b. Cell destruction
c. Sweating
d. None of the above

160. Bluish green colour of some highly coloured egg shell is due to pigment

a. Oocyan
b. Metalin
c. Phsophovyrin
d. None of the above

161. ___________ is transparent protein coating on the surface of shell

a. Cuticle
b. Shell membrane
c. Vitelline membrane
d. All of the above

162. Name the emulsifying components of egg yolk

a. Lecithin
b. Cholesterol
c. Lipoproteins
d. All of the above

163. Phospholipid fractions of yolk constitute essentially of

a. Phosphotidyl cholin
b. Phosphotidyl ethanolamine
c. Both of the above
d. None of the above

164. Ovoinhibitor is capable of inhibiting

a. Trypsin
b. Chymotrypsin
c. Fungal and bacterial proteases
d. All of the above

165. Ovomucoid is a glycoprotein containing

a. Galactose
b. Mannose
c. Both of the above
d. None of the above

166. Freeze drying of poultry meat has the advantage of

a. No oxidative rancidity
b. No foaming
c. No case hardening
d. All of the above

167. Light weight laminated pouches made of____________ are quite suitable for the packaging of freeze dried meat.

a. Aluminium
b. Vinyl
c. Polyester
d. All of the above

168. Suggest the additives which are capable to prevent the oxidative rancidity of mechanically deboned poultry meat

a. Polyphosphates
b. Citric acid
c. Rose mary spice
d. All of the above

169. Sweet pickle usually consists of ______________dissolved in water

a. Salt
b. Sugar
c. Salt petre
d. All of the above

170. Sterility of canned chicken meat can be checked by incubating cans at

a. 35°C for 10 days
b. 30°C during 5 days
c. Both of the above
d. None of the above

171. To disperse chlortetracycline uniformly in ice___________is added to the water before freezing

a. 100ppm of CMC
b. 300ppm of Hacl
c. Both of the above
d. None of the above

172. Oxidative rancidity in poultry meat is caused by oxygen and is catalysed by

a. Metal and light
b. Metal and heat
c. Metal and oil
d. None of the above

173. Oxidative rancidity in poultry meat is caused by unsaturated fat and is catalysed by

a. Ultraviolet light
b. Heat & high pressure
c. Metal and oxygen
d. All of the above

174. Plastic films commonly used for the packaging of fresh poultry are

a. Flexible packaging film
b. Overwrap
c. Composite layer containing paper& foil
d. All of the above

175. Egg shell matrix is made up________________protein.

a. Glycoproteins
b. Proteoglycans
c. Both of the above
d. All of the above

176. Frying of eggs may cause

a. 25% loss of thiamin
b. 25% loss of vitamin B_{12}
c. 31% loss of folacin
d. All of the above

177. Liquid egg is stabilized by

a. Glycerin
b. Sodium meta phosphate
c. Sugar and salt
d. All of the above

178. Write the modes of microbial penetration in egg shell

a. Colonization
b. Penetration
c. Both of the above
d. All of the above

179. Why singeing of poultry carcass is done over blue flame for 1-5 seconds

a. To remove filoplumes
b. To remove feathers
c. To remove skin
d. All of the above

180. The albumen enzyme which has a lytic action on bacterial cell wall is

a. Lysozyme
b. Protease
c. Catalase
d. None of the above

181. Poultry and fish meat is considered as

a. White meat
b. Red meat
c. Pink meat
d. None of the above

182. India ranks in global fish production.

a. First
b. Second
c. Third
d. Fourth

183. Eggs with mixed rots is termed as

a. Addled eggs
b. Adulterated eggs
c. Adurant eggs
d. None of the above

184. Technical albumen is used in

a. Medicines
b. Adhesives
c. Media
d. None of the above

185. Vitelline membrane is found in the __________ portion of egg

a. Egg albumen
b. Egg yolk
c. Shell membrane
d. None of the above

Q.2 Fill in the blanks

1. Egg albumen provides an unfavourable medium for microbial growth due to presence of____________________.
2. Biological value of egg protein is considered as ____________________.
3. ____________________protein in egg albumen is potent binder of riboflavin.
4. The avian vitelline membrane (VM) is a multilayered proteinaceous structure separating egg white from ____________________.
5. Shell comprises around ____________ percent of the total egg weight.
6. Water content of poultry meat ranges from _______________percent.
7. Thinning of egg white on prolonged storage is due to ______________.
8. Air sac is usually present on the broader end of egg because __________.
9. Pseudo H-zone is the least dense region in A-band explain ___________.
10. Quick freezing is preferred over fast freezing because _______________.
11. Elastin is insoluble in nature due to ____________________.
12. Rigor mortis is completed early in PSE meat due to _______________.
13. DFD meat is more liable to spoilage by Micro-organisms because of __________.
14. Refrigeration do not prolong the shelf life of meat significantly because ________________.
15. Hot deboning of carcass is not recommended because________________.
16. Irradiation is also referred as cold sterilization because ________________.
17. Yolk constitutes around____________________percent of total egg weight.
18. Poultry and fish are generally chilled by immersion in __________________.
19. The absorption of meat protein is about____________________ Percent.
20. It is generally accepted that evolution of birds was from______________.
21. All ducks except Mascovy's are descendent from ________________.
22. Egg yolk contains excellent source of Lutein & zea-xanthin which are responsible for____________________.
23. ____________________is a toxic non-protein amino acid structurally similar to tyrosine.
24. Choline is found in egg in the form of ____________________.

25. The contents of cholesterol in egg may be reduced through the supplementation of ____________________.
26. ____________________ is the most preferred oil in the preservation of shell egg by oil coating.
27. ____________________ is a antioxidant most commonly preferred in the preservation of poultry pickle.
28. If dressed chicken is free from pinfeathers then it will come under grade ____________.
29. ____________________ is the most preferred time for catching the poultry birds for slaughter.
30. Water content in the fish meat is ____________________percent.
31. The most preferred preservative medium for fish is____________________.
32. Singeing of poultry carcass is done over blue flame for____________ seconds.
33. In poultry grading system kind refers to ____________________.
34. Egg mélange is pasteurized at____________________°C for__________ minutes to kill pathogenic organisms.
35. Among all the amino acids egg shell waste with adhering albumen is rich in____________________.
36. While preparing turkey salami the moisture to protein ratio should not exceed____________.
37. Leg muscle of poultry grows at ______________rate than breast muscle.
38. Egg powder is higly hygroscopic and liable for ______________.
39. The pH of old stored eggs is ______________as compared fresh eggs.
40. Birds showing symptoms of infectious laryngotreachitis in an advanced stage with cyanosis are ______________ for slaughter for food.
41. Weight loss (shrinkage) of broilers between farm and processing plant is ____________.
42. ____________________ of poultry birds leads to poor keeping quality.
43. To chill dressed bird in water, crushed ice is used to an internal temperature of ____________________.
44. Yolk is formed largely during the final ______________ days prior to laying.

45. At the time of hatching, the ovary of female chicken has around ____________ ova present.
46. Shell thickness should be least ______________ for better marketing handling.
47. Roche colour fan used for measuring yolk colour consists of____________ coloured plastic tube.
48. Relative humidity recommended for holding eggs between ____________.
49. Temperature less than ________________ in farm cooling rooms causes sweating.
50. Raw egg white contains _________________ which is harmful to human health. This can be avoided by boiling or heating the eggs.
51. Commercial egg yolk contains ______________ egg white.
52. Maximum reducing sugar in spray dried egg white is__________________.
53. Duck egg white coagulates when held at 55^{0}C for____________________.
54. Commercial liquid yolk has about__________________ solids.
55. Fresh egg yolk has about _________________ solids.
56. Loss of ____________from albumen results in rise in pH from 7.6 to 8.9 during ageing.
57. During pickling maximum acid concentration is achieved in _______________day in white and in about six days in the yolk.
58. Scotch eggs are deep fat fried in oil at ___________________.
59. Pancakes are______________than crepes so required longer cooking time.
60. Lysozyme makes up about ________________ of the egg albumen.
61. Frozen yolk is produced by adding_____________salt or sugar to improve viscosity of thawed products.
62. Loss of weight during cooking poultry averages about ______________.
63. Loss of weight during roasting of poultry is ___________________.
64. During preparation of an omlet________________properties of eggs proteins play an important role.
65. Entire cholesterol in the egg is concentrated in __________________.
66. For cake making ________________provides better foaming properties than whole egg liquids.

67. Three months storage of frozen egg improve_______ quality of whole egg.
68. ____________________ egg protein exhibits anti-trypsin activity.
69. The pH of albumen from a freshly laid egg is between ____________________.
70. During storage of shell eggs pH of albumen increases to a maximum value of __________.
71. ________________ are common additives added to frozen egg products at levels of about 10% to control gelation of products containing egg yolk.
72. Egg is deficient in vitamin________ and __________ minerals.
73. ________________ enzyme is used for desugarization of eggs, which catalyses the conversion of glucose to gluconic acid.
74. Shell eggs should be cooled to __________ ^{0}C when kept in the cooling rooms.
75. Greenish black discolouration in hard boiled eggs is due to interaction of ________________compounds.
76. Aged eggs are ______________ to peel on boiling.
77. Suitable humidity in egg holding storage room is desirable to prevent loss of water which causes__________________ in air cell size.
78. Spoilage of shell egg is usually caused by ____________________.
79. The albumen protein __________________ complexes biotin.
80. The pH of egg protein __________________ during storage.
81. There is a chemical resistance to micro-organisms due to _______________ in egg albumen.
82. Fat content of whole egg is __________________ as compared to egg white.
83. Egg is a ________________ source of vitamin-c.
84. __________________properties of egg proteins play an important role in omlet preparation.
85. Egg yolk has __________________ water than whole egg.
86. Albumen index of egg usually varies between __________________.
87. ____________________ are added to prevent gelation of frozen egg pulp.
88. ________________ spp of mold grows only on the surface of egg shell.
89. Normally chalaza at the large end of the egg has an _________________ wise twist.

90. The outer shell membrane is ________________ times as thick as inner one.
91. Egg powder is packaged in tin containers with an ____________________ gas.
92. ________________________ properties of egg albumen are used in cake baking.
93. Eggs are pasteurized primarily to destroy bacteria pathogenic to humans, especially __________________ spp.
94. Eggs act as a ______________________ agent in baked foods.
95. The thick layer immediately surrounding the yolk is called the ____________.
96. Giblets include ____________________________________.
97. _______________________ appear to be most active antioxidant compounds in smoke.
98. ___________________ is the principal bactericidal agent in the smoke used for smoking poultry.
99. Modified kosher killed birds are those where the jugular vein is severed just below the jowls so that the windpipe and ____________________ remain uncut.
100. _____________________ favours formation is an oil-in-water emulsion.
101. Cholesterol tends to form a ______________________.
102. All riboflavin in egg albumen is bound as ___________________.
103. Yolk lipid contains ____________________________.
104. Plasma is composed of ________________ which are lipid –free globular proteins and low density lipoproteins.
105. The amount of carbohydrate in yolk may be as high as _____________ percent.
106. Avidin binds with ______________________.
107. Conalbumen complexes with ___________________.
108. Water foil is hard scalded at 160-180^{0}F for __________ seconds to release the feathers.
109. Bone darkening can be prevented by storing birds at a temperature below_____________.
110. Salt, Sugar, Sodium Nitrate/Nitrite are ___________________________.

111. Ascorbic acid, polyphosphate and MSG are ____________________.
112. Curing solution can be applied by ____________________.
113. To produce ready to eat smoked poultry, the internal temperature of the meat should reach ____________________.
114. The percentage of bone solids in final mechanically deboned poultry meat must be below ____________________ percent.
115. The frying fats for poultry meat should contain ____________________ agents to retard the breakdown of fats into free fatty acids.
116. One detergent for use of egg shell is the higher proportion of ____________ in relation to other nutrients.
117. To prevent gelatin in frozen yolk, ____________________ upto a level of 10% are added.
118. ____________________ in egg holding room is desirable to prevent loss of water which causes increase in air cell size.
119. Bacterial load of egg powder goes on ____________________ during storage.
120. Spoilage of shell egg is usually caused by ____________________ bacteria.
121. Canned poultry meat is generally spoiled by ____________________ bacteria.
122. The pH of egg ____________________ during storage.
123. An ____________________ is a preferred method to stun poultry prior to slaughter.
124. There is chemical resistance to micro-organism due to ____________ in egg albumen.
125. Poultry meat is a good source of Vitamin ____________.
126. Fat content of whole egg is ____________ as compared to egg white.
127. ____________________ practices have an effect on the carcass yield of poultry.
128. ____________________ is used for wet chilling and cooling for dry chlling of dressed poultry.
129. In scientific method of slaughter jugular vein and carotid artery are cut just below the ____________________.
130. Mature birds need ____________________ scalding.

131. Calcium content of mechanically deboned poultry meat is ____________ than that of hand deboned.

132. Raw eggs are ______________________ to digest.

133. Egg is a poor source of vitamin________________.

134. Polyphosphates are added to chilling water to reduce ___________________ losses.

135. Fabrication of ready to cook poultry rolls should be such that the items should not break into more than _________________ when sliced warm after cooking.

136. Ready to cook poultry with one disjointed and one protruding broken bone can be placed under ________________ of USDA grade.

137. Ready to cook poultry with misshaped legs can be classed under USDA ______________.

138. 'Kind' legs is a ready to cook poultry part that includes ________________.

139. Natural proportion of skin in chicken shall be ___________ in raw and ___________ when cooked.

140. Wing drummate consists of ________________ portion of poultry with adhering skin and meat.

141. Cooling and retort water used for canning of poultry should not contain more than __________________ of zinc oxide.

142. Boneless turkey breasts should not have more than _______________ skin or raw basis.

143. Minimum percentage of cooked deboned meat in poultry ravioli is _______.

144. Minimum percentage of deboned meat in cooked poultry Tetrazzini is ______________.

145. Chicken Fricassee contains _______ cooked deboned chicken, ___________ peas and __________________ sauce

146. Ready to cook chicken cooked in moist heat with barbecue sauce is known as ______________________.

147. Boned poultry, solid packed should yield _________________ poultry meat on cooked basis.

148. In USA all liquid eggs should be pasteurized at _________________ minutes before converting it into a dehydrated product.

149. Four hundred forty four grams of dry yolk is equal to _______________ fresh yolk (of large sized eggs).

150. Hard chilled poultry is usually held in frozen storage for not more than _______________days.

151. Egg white has _______________ ovalbumen.

152. Egg yolk begin to coagulate at _______________ ^{0}C temperature.

153. Sugar _____________ the temperature required to cause coagulation in proportion to the sound added.

154. Ozone in chilling plant will _________________ mould and yeasts.

155. Slow freezing _________________ the structural changes in muscles.

156. In blast freezing air moves at the rate of ______________________.

157. Poultry processing plants with flow away system requires__________________ litres of water for each 1000 birds.

158. Poultry processing plants with non flew away systems require _______________ litres of water for each 1000 birds.

159. The recommended lightening in poultry processing plant is ___________ candles in inspection areas

160. The recommended lightening in poultry processing plant is ____________ candles in other working areas.

161. Low humidity during smoking causes excessive ____________ and shrinkage.

162. For rodent control in slaughter houses all feed stuff should be stored at least 18 inches away from the walls and _______________ inches above the floor.

163. In poultry carcasses most bruises occur within _______________ hrs prior to processing.

164. Egg____________________ is used as a meat binder.

165. Egg______________is used in soups and salad dressing.

166. Addition of __________________improves stability of baked unsweetened plain whole egg.

167. Poultry birds are the largest single source of pathogenic micro-organisms called _________________ in which causes digestive disturbances in human beings.

168. For controlling rodents in poultry processing plant, __________________ are the commonly used methods.

169. For electric stunning of poultry, a current of ________________ seconds is give to the birds.

170. To produce ready-to-eat smoked poultry products, the internal temperature of meat during smoking reach ____________ °F.

171. Maximum shelf life of smoked poultry meat at 0^0F is ____________ months.

172. The zone of maximum ice-crystal formation ranges from ____________ ^{0}F.

173. ____________________ are the plastic films commonly used for packaging of fresh poultry.

174. ________________ is the ratio of vapour pressure of water in foods to that of vapour pressure of pure water.

175. As a rule ______________ times as much air is required to heat food as it needed to carry the moisture vapour from food.

176. ____________________ are only freeze dried products which have attained commercial success.

177. Radiation doses are usually expressed in 'rad' which corresponds to energy absorption of _______________ ergs per gram of material.

178. ________________ is the pigment responsible for the desirable red colour of fresh poultry meat

179. The liquid eggs are pasteurized at ____________ minutes to kill Salmonella organism for proportion of egg powder.

180. _________________ is emulsified product produced using egg yolk, fat and water.

181. The rigor mortis is completed in ____________ hrs in poultry birds after slaughter.

182. ____________________ protein in egg has antibacterial properties.

183. __________________ is an example of putrefactive anaerobe bacteria important in the thermal process time evaluation canned poultry products.

184. The moisture content of spray dried egg powder is reduced from ____________ percent in a vacuum shelf drier for **3** hrs.

185. The temperature of water for semi scalding of poultry is ________________

186. The temperature of water for sub-scalding of poultry is ___________ respectively.

187. Microbial and enzymatic desugaring of eggs is carried out by adding ____________________ respectively.

188. The inner shell membrane of egg is ________________ times.

189. Liquid whole egg is stabilized by adding ________________ for frozen egg preparation.

190. ____________________ is a protein responsible for gel structure of thick albumen.

191. Mositure content of whole egg is _____________ while that of albumen is _________.

192. γ-benzpyrene is a carcinogenic compound formed during the smoking of poultry meat at a temperature higher than ________________ ^{0}F.

193. The egg stays in the uterus and vagina of hen for ______________ hours.

194. Egg is deficient in __________________________ vitamin.

195. Cholestrol forms _______________ of yolk lipid.

196. ________________ egg white protein is a typsin inhibitor.

197. The temperature of egg washing water should be ________________ above the temperature of egg contents.

198. ________________ ingredient in poultry feed causes pink dicolouration.

199. Yellow colour of yolk is mainly due to the ____________________ pigment.

200. __________________ compound in wood smoke imparts anti-oxidant properties during the smoking of poultry meat.

201. Whole, halves or quarters of young chickens can be cooked by broiling at _____________ ^{0}F.

202. Sausages may be cooked in water by simmering at ____________________.

203. Sharp freezing consists of freezing at _____________________ or lower without mechanically agitating the air.

204. The temperature range between _______________ is the zone of maximum crystal formation when poultry meat is frozen slowly.

205. Well packaged turkeys and cut up chicken stored at ____________ will remain in good condition for one year.

206. _______________ method of sticking poultry keep the trachea and oesophagus to remain uncut.

207. Semi-scalding is carried out at __________________________.

208. The poultry carcasses must be chilled to at least ____________ internal temperature.

209. The smoking is continued until a minimum internal temperature of________ is reached.
210. Insecticides containing ________________ produces immediate response to kill insects.
211. The zoological name of oriental cockroach is ____________________.
212. Emulsifying and water holding capacities of deboned poultry can be ___________ by removing the skin from the mechanically deboned poultry (broiler) backs.
213. The picking and pinning feathers, the poultry carcasses singed to remove hair like appendages is called __________________.
214. Canning of whole chicken an autoclaving temperature of________________ minutes produces acceptable products.
215. Hydrates such as sucrose, glucose have been found to heat denaturation by preventing the formation of _______________ groups.
216. _________________ is most by used for water fowl which are immersed in hot water (71-72^{0}C for 30-60 seconds).
217. Loss of baking quality of whole egg powder is associated with____________.
218. Regulations specify poultry carcasses to be chilled to _______________.
219. Processed MDPM usually have a TBA value of ____________________.
220. Barbecuing chicken grills can be made ___________________________.
221. Off flavour develop in whole egg powder is associated with __________.
222. Particular strain of bacteria is quite resistant to heat pasteurization of egg and egg products is _________________________.
223. The absorption of meat protein is about____________________Percent.
224. It is generally accepted that evolution of birds was from _____________.
225. All ducks except Mascovy's are descendent from _________________.
226. Egg yolk contains excellent source of Lutein & zea-xanthin which are responsible for____.
227. _______________is a toxic non-protein amino acid structurally similar to tyrosine.
228. Choline is found in egg in the form of ____________________________.
229. The contents of cholesterol in egg may be reduced through the supplementation of _______________________.

230. ________________ is the most preferred oil in the preservation of shell egg by oil coating.

231. ______________ is an antioxidant most commonly preferred in the preservation of poultry pickle.

232. If dressed chicken is free from pinfeathers then it will come under grade ____________________.

233. ____________________ is the most preferred time for catching the poultry birds for slaughter.

234. Water content in the fish meat is ________________percent.

235. The most preferred preservative medium for fish is ________________.

Q.3. Write true or false against each statement

1.	Most fat in poultry meat remains confined underneath the skin.	( )
2.	Broilers are generally caught at day time under good light.	( )
3.	Singeing is generally performed over red flame to remove filoplumes .	( )
4.	Egg white or albumin provides an unfavourable medium for microbial growth.	( )
5.	Poultry carcass showing post mortem lesions of tuberculosis shall be passed for human consumption.	( )
6.	Poultry carcass showing signs of Ranikhet disease is totally rejected for human consumption.	( )
7.	Carcass exhibited signs of Marek's disease shall be passed fit for human consumption.	( )
8.	Turkeys are scalded at 60^{0}C for 60seconds.	( )
9.	Gel like structure of thick egg white is due to ovomucoid.	( )
10.	Singeing is performed in turkey slaughter only.	( )
11.	India is the second largest producer of the food.	( )
12.	Poultry term comes from the English word Poult.	( )
13.	Broilers with about 750 g dressed weight are preferred for barbecuing	( )
14.	Poultry carcass require sufficient amount of oxygen in its packaging.	( )
15.	Surimi is a poultry product.	( )
16.	EIC is the chief enforcement body for export	( )
17.	For the cold storage of eggs, relative humidity in cold storage should be 80-90%.	( )

18.	Conalbumin of egg is responsible for chelation of iron, zinc and copper.	( )
19.	Mature poultry is generally considered better for canning than younger birds.	( )
20.	India ranks fourth in global fish production.	( )
21.	Dressing percentage of poultry is comparatively higher as compared to other animal species.	( )
22.	Spoilage of egg is not caused by molds but by bacteria.	( )
23.	The term 'Giblet' includes liver, heart, neck and gizzards.	( )
24.	Ovomucin is a tripsin inhibitor protein present in the albumin of egg.	( )
25.	Bone content of eviscerated male turkey carcasses are usually less than their female counterparts of the same age, breed and strain.	( )
26.	Over fatness in meat type chicken cannot be checked by hypothalamic control of appetite.	( )
27.	It has been observed that there is no positive effect of high protein levels on breast meat yield.	( )
28.	Lysine arginine antagonism is a common problem in the growth of chicken as well as turkeys.	( )
29.	Pseudomonas organisms will grow at water activity level 0.97 and Coliforms will not.	( )
30.	Turkey hams after cooking/smoking are cooled to an internal temperature of 4.2^{0}C.	( )
31.	Sex has no marked effect on weight and composition of poultry carcass yield.	( )
32.	In poultry meat sticks not more than 10% extenders and 30% breading are permitted.	( )
33.	Change in colour is a major problem of poultry products.	( )
34.	Poultry fat contains lower amounts and unsaturated fatty acids as compared to pork.	( )
35.	Sodium hypochlorite solution may be used at 40^{0}C for washing dirty eggs.	( )
36.	Scalding time and temperature have no effect on keeping quality of dressed chicken.	( )
37.	Oxidative rancidity may be reduced by using cryovac for frozen birds.	( )
38.	Singeing is done after evisceration.	( )
39.	Dressing percentage is higher in layer than broiler.	( )
40.	The protein content of poultry meat is about 18%.	( )

41.	Broilers have flexible breast and bone cartilage.	( )
42.	Enzyme most commonly used in tenderizing meat is papain.	( )
43.	The normal cooking time for Tandoori chicken in earthen tandoor is about 1 hour.	( )
44.	Cooking in liquid medium is called stewing.	( )
45.	The shelf life of dressed poultry stored at 4^0C is about 3 months.	( )
46.	The protein fibers of the mamillary matrix region are not interconnected with the outer shell membrane.	( )
47.	Crystals within the calcified matrix have their long axis oriented towards the shell surface.	( )
48.	Matrix fibers do not pass through calcite crystals but simply surrounds them.	( )
49.	Fore canals are funnel shaped and are at right angles to shell surface.	( )
50.	Maximum diameter of fiber in inner shell membrane is around 1.5 microns.	( )
51.	There are about 7000 to 17000 pores in the chicken egg shell.	( )
52.	Compared to cuticle protein shell membrane have a lower content of histidine, cystine and proline.	( )
53.	Total solid content of albumen is independent of the strain and age of hen.	( )
54.	Protein content of egg albumen increases by 0.09 g for each increase of one gram egg weight.	( )
55.	Lipid content of Leghorn hen eggs are around 40.7%.	( )
56.	Eggs are mostly used in cake preparations because it produces large volume of stable foams which coagulate during heating.	( )
57.	Ready to cook carcass yield of broilers as a percentage of live weight is 76-78%.	( )
58.	Hard scalding is done for ducks and older birds at 71-82^0C for 30 seconds.	( )
59.	Egg yolk is a excellent emulsifier in fats or oils and water.	( )
60.	Two point suspension is used in broiler carcasses by hanging both hock shackles for evisceration.	( )
61.	The amount of moisture absorbed by poultry carcasses ranges from 2-3.25%.	( )
62.	Ovomucoid is not coagulable by heat.	( )
63.	USDA regulations specify that poultry carcasses must be chilled to at least 4.4^0C.	( )
64.	Sub-scalding is done at 58-60^0C for 30-75 seconds.	( )

65.	To prevent major effect on weight loss in eggs during storage, a humidity of 75-80% is recommended to store eggs.	( )
66.	If liquid eggs are not desugared prior to dehydration , brown discolouration occurs in the egg powder due to reaction between lipids and carbohydrates.	( )
67.	The emulsification capacity of egg yolk is attributed to its triglycerides content.	( )
68.	Egg albumen protein ovomucin is mainly responsible for the foaming power of egg albumen.	( )
69.	Egg yolks with deep yellow colour is higher in nutritive value than those with a lighter colour yolks.	( )
70.	Eggs in cold storage should be kept at a temperature of 1-2^{0}C and RH of room should be atleast 90%.	( )
71.	Eggs graded as AA grade should only be used for the production of dehydrated egg products.	( )
72.	The pH of egg albumen increases during storage due to escape of H_2S through the pores of the shell.	( )
73.	Egg wash water temperature should be at least 11^{0}C higher than egg temperature. The temperature differences should not exceed 10-11°C.	( )
74.	Sodium chloride is added to frozen and dehydrated egg products as whipping aids.	( )
75.	Pasteurization temperature of liquid egg should be sufficient to destroy Mycobacterium tuberculosis.	( )
76.	Per capita egg consumption in India has increased to 50eggs per anum.	( )
77.	Raw eggs are easy to digest.	( )
78.	Egg albumen contains all fat soluble vitamins.	( )
79.	Bacterial load of egg goes on increasing during storage.	( )
80.	Moisture content of whole eggs is 65-70%.	( )
81.	Air cell size of 3-6 mm means A grade eggs.	( )
82.	Pink rot in egg is caused by mold growth.	( )
83.	Egg white is the only natural barrier against micro-organisms.	( )
84.	For the manufacture of egg powder the egg liquids are pasteurized at 62.5^{0}C for 3.5 minutes.	( )
85.	Fresh eggs are less susceptible to bacterial spoilage.	( )
86.	Ideal temperature for oil treatment of eggs is 120^{0}C.	( )
87.	Moisture content of egg powder should be 10% for safe storage.	( )
88.	Thermostabilization of eggs is done at 140^{0}F for 5 minutes.	( )

89.	Technical grade albumen is prepared from fresh eggs.	( )
90.	Actin and myosin are proteins of egg albumen.	( )
91.	The egg shell consists of around 94-97 percent calcium carbonate.	( )
92.	The germinal disc on the surface of egg yolk is always present whether the egg is fertilized or not.	( )
93.	The egg yolk colour is unrelated to nutritional value, it only reflect the amount of xanthophyl pigment in the hen's diet.	( )
94.	Albumen contains more lipids and less moisture than yolk.	( )
95.	Egg protein is one of the highest quality proteins known for human food.	( )
96.	Egg yolk contains no natural emulsifiers.	( )
97.	Pink albumen can be caused by cottonseed and soybean meal in hen's diet.	( )
98.	In general, shrinkage is greater in older or heavier birds than in young birds and light weight ones.	( )
99.	In ducks, among the cut-up parts, legs and thighs produce highest cut percentage based on carcass weight.	( )
100.	Nutrient losses of poultry meat from freezing and thawing are primarily due to drip.	( )
101.	In poultry, white meat contains less protein than dark meat.	( )
102.	Salt in cure mixture influences the flavour and texture of meat and acts as a preservative.	( )
103.	Canned chicken usually contains a chicken with chicken broth but without giblets.	( )
104.	The faster the poultry carcasses pass through the zone of maximum crystal formation, the larger crystals in the frozen product.	( )
105.	Approximately 10 kg of liquid albumen are required to produce 1 kg of dried egg white.	( )
106.	Shelf life of stabilized egg yolk is not more than 8 months under refrigeration.	( )
107.	Human system cannot digest the avidin-biotin complex present in raw egg.	( )
108.	Co_2 is liberated from fertile eggs only.	( )
109.	Whole egg products do not have good whipping ability by themselves.	( )
110.	It is always desirable to wash all the eggs before releasing into the marketing channel.	( )
111.	Albumen proteins are mostly conjugated proteins.	( )

112.	Frequent collection of eggs from the pen is deleterious from quality point of view.	( )
113.	Nutrients are distributed equally in albumen and yolk.	( )
114.	The thickening property of egg albumen on heating has best been demonstrated in puddings and candles.	( )
115.	Egg is a moderately good source of ascorbic acid for babies.	( )
116.	Powdered whole egg has the largest uses in angle of food cakes.	( )
117.	Dried yolk products are heat-pasteurized in the liquid state while egg white products are heat treated after drying.	( )
118.	Egg white is pre-treated with acids before drying for better desugaring.	( )
119.	Electron microscopy studies of egg shell indicate that the matrix fibres pass through the calcite crystals rather than surrounding them.	( )
120.	Ovomucoid is a heat resistant glycoprotein.	( )
121.	The ovomucin-lysozyme complex plays an important role during storage.	( )
122.	Conalbumen is a phosphoglycoprotein.	( )
123.	Heat denaturation of albumen is more in alkaline pH than neutral.	( )
124.	In natural state about 3 moles of biotin are bound to one mole of avidin.	( )
125.	Pores are evenly distributed throughout the shell surface.	( )
126.	Ovoglobulin is an excellent foaming agent.	( )
127.	Nutritive value of frozen and dried egg is inferior to that of fresh egg.	( )
128.	The emulsifying ability of albumen is considerably low as compared to yolk.	( )
129.	Phospholipids of yolk consists mostly phosphatidyl ethanolamines.	( )
130.	The loss of foaming power of whole egg powder can be prevented by addition of carbohydrates to the mélange.	( )
131.	Ovalbumen is converted to S-ovalbumen during storage.	( )
132.	Ovomucoid is a trypsin-inhibitor.	( )
133.	Cold desugaring is an ideal method for stabilization of egg mélange.	( )
134.	Egg albumen contains two times as much water as yolk.	( )
135.	Flattening of yolk is caused due to escape of water from albumen into it.	( )
136.	Egg yolk is prescribed for babies because of low fat content.	( )

137.	Liquefaction of egg white is hastened at an elevated temperature.	( )
138.	Conalbumen is more heat-sensitive but less susceptible to surface denaturation than ovalbumen.	( )
139.	Lecithin is the only constituent responsible for emulsifying property in yolk.	( )
140.	Double yolked eggs are nutritionally poor.	( )
141.	Ovalbumen is phosphoglycoprotein with isoelctric point of 4.6.	( )
142.	Anaesthetizing the poultry birds increases the feather pulling forces by 25-50%.	( )
143.	Slack scalding is carried out at 123-130^{0}F for 90-120 seconds.	( )
144.	The mechanically deboned poultry meat has a good shelf life in cool storage as compared to whole carcass.	( )
145.	The galvanized grills are quite suitable for barbequing poultry meat.	( )
146.	The secret of tender juicy barbecued poultry is fast cooking at a higher temperature.	( )
147.	Oil in egg yolk and protein in albumen make egg a desirable material for tanning leather.	( )
148.	Lactic acid is added to prevent damage by heat to the egg white during dehydration.	( )
149.	Egg mélange is pasteurized at 140^{0}C for 3.5 minutes to kill Salmonella spp.	( )
150.	About one mega rad is required as sterilization dosage for chicken meat.	( )
151.	Poultry carcass having too much fat results in high shrinkage during smoking.	( )
152.	Permanent hardness of water is caused by the presence of calcium and magnesium sulfate and chlorides.	( )
153.	In the inspection area, atleast 10 foot candles are recommended for lighting.	( )
154.	50ppm chlorine solution is commonly used for sanitizing the poultry processing equipment.	( )
155.	Cationic surface active agents move the whipping ability of dried egg products.	( )
156.	Pink rot in eggs is caused by mold growth.	( )
157.	Ascorbic acid and polyphosphates act as curing adjunct.	( )
158.	Egg yolk contains all fat soluble vitamins.	( )
159.	For the manufacture of egg powder, egg liquids are pasteurized at 100^{0}C.	( )

160.	There is some weight gain during dry chilling of dressed carcass.	( )
161.	Crop is part of giblet.	( )
162.	Albumen quality decreases with storage of egg.	( )
163.	Moisture content of egg powder should be 10% for safe storage.	( )
164.	Chilled poultry carcasses should be packed in 800 gauges LDPE for economical marketing.	( )
165.	Frankfurters are emulsion type sausage.	( )
166.	Bleeding time in poultry is 10 minutes.	( )
167.	Salt is added last to chicken meat in emulsion type sausages.	( )
168.	Detergents to be used in poultry canning plants for sanitation purpose should not have active alkalinity of 500 to 900ppm, where hands come in contact.	( )
169.	Corrosiveness of trisodium orthophosphate is less than sodium metasilicate used as a detergent for sanitization.	( )
170.	The combined residual chlorine of water of poultry processing plant is total residual chlorine minus free residual chlorine.	( )
171.	Chlorination of water supply beyond the break point is known as in-plant chlorination.	( )
172.	Lagooning adopted for biological disposal of waste water from poultry processing plants consist of only four principles such as biological action, sedimentation, soil adsorption and evaporation.	( )
173.	Turkey wings from around 7-8% of chilled ready to cook weight.	( )
174.	Ready to cook poultry with scattered no protruding pins and hairs are classed under'B'grade of USDA standards.	( )
175.	Evisceration techniques have no effect on water uptake of dressed carcasses.	( )
176.	Heme ferroprotoporphyrin of myoglobin combines with histidine but not lysine.	( )
177.	Motner of pearl effect in dressed poultry can occur even if they are properly dressed and packed.	( )
178.	S. senftenberg is quit resistant to the pasteurizing effect of egg liquids.	( )
179.	Sugar used in curing poultry not only adds to its flavour but also conserves its protein.	( )
180.	Hard wood smoke contains phenol to an extent of 10 ppm.	( )
181.	Aldehydes present in smoke are responsible for the glossy finish of smoked poultry.	( )

182.	When poultry meat is ground hot and the cure to it is added, during grinding the extractable salt soluble protein is low.	( )
183.	In emulsion type meat sausage the original meat colour gets diluted.	( )
184.	Fungal protease is usually obtained from *Aspergillus oryza* which in turn is only used for tenderizing poultry meat.	( )
185.	Liquid egg white when concentrated to have solid contents of 15% and above, it becomes very difficult to pump for drying.	( )
186.	Poly-phosphates have been shown to have beneficial effect on the beating property of whole egg solids.	( )
187.	Salted yolk acidified to pH 6.2 and then pasteurized at 140^0F suffers from loss in emulsifying capacity.	( )
188.	More than 250 chemical compounds have been identified to be involved in production of flavour in poultry meat.	( )
189.	Raw broiler has very good flavour which is lost on cooking.	( )
190.	Air super saturated and moving at high speed will prevent desiccation and remove heat of the carcass.	( )
191.	If the temperature difference between meat surface and air becomes small air speed must be increased to avoid desiccation.	( )
192.	Not all the muscles are free to cold shortening.	( )
193.	Freezing and storing of raw egg yolk below -6^0C leads to gelation.	( )
194.	The chalazae of egg are low in mucin content.	( )
195.	Removal of naturally occurring salts from egg white impairs its ability to coagulate.	( )
196.	Egg white devoid of globulin and ovomucin will form foam quickly.	( )
197.	Egg yolk is a good emulsifying agent.	( )
198.	Young birds are considered better for canning than mature poultry.	( )
199.	Poultry meat subjected to prolonged curing and smoking is difficult to distinguish from smoked pork or beef products.	( )
200.	The type of wood chip or saw dust used for smoking will not influence the flavour of the finished product.	( )
201.	Crystalline is a purified protein derived from egg white.	( )
202.	Pentrometer is used to measure the foaming property of egg.	( )
203.	Water hardness refers to the amount of calcium and manganese salts present in water.	( )
204.	The water used in slaughter houses Containing 2 Coliform organism per 100 ml is permitted.	( )

205.	Excess of pH or water can cause corrosion of water pipes.	( )
206.	Wetting agents added to scald water should not exceed two percent according to USDA.	( )
207.	In poultry slaughter houses of 10 thousand capacity per day, for appropriate and economic cleaning arrangement at least a team of 5 man be provided.	( )
208.	The efficacy of electrical stunning is dependent on the individual of faster of voltage, amperage or time.	( )
209.	Tranquilizers in most cases increases the force required for removal of feathers.	( )
210.	Singeing is made to remove the filoplumes.	( )
211.	The bacterial count on equipments and working surfaces is reduced by as much as 88% by chlorine treatment.	( )
212.	During transport of poultry maximum permissible level of shrinkage is 6%.	( )
213.	True Kosher slaughtering comes under Jewish method of slaughter.	( )
214.	Hard scalding is desirable for water fowl as well as for Geese.	( )
215.	For efficient removal of oil glands the carcasses be suspended by the neck rather than the hocks.	( )
216.	Turkeys are usually suspended by three point suspension.	( )
217.	Pericardial sac of heart is included in giblet.	( )
218.	Bulk containers used for packaging held 68-70 lb of chicken and 20-30lbs of ice.	( )
219.	The most effective component against microorganisms produced from wood smoke is phenol.	( )
220.	Smoke free soft wood produces light pink to mahogany colour.	( )
221.	Turkeys used for smoking should be under 2 yrs of age.	( )
222.	Giblets are mostly cured and smoked.	( )
223.	According to ISI, yeast and mold count per gram of spray dried egg powder should be 100.	( )
224.	The mixture of egg yolk and albumen form soft coagulam at 75^{0}C	( )
225.	Desugaring of the whole egg liquid is made by 0.5% yeast fermentation of 30^{0}C for 1.5 hrs.	( )
226.	The effectiveness of pasteurization of egg is tested by alpha amylase activity of egg yolk test.	( )
227.	Whole egg are cleaned by treating with 2% sodium hypochloride solution at 40^{0}C water temperature for 3 minutes.	( )

228.	Poultry patties and nuggets are now found on fast food menus as an alternative to hamburger.	( )
229.	Chicken sandwitch open and class should contain atleats 59% and 35% chicken meat respectively.	( )
230.	Chicken chilli contain atleast 28% chicken meat.	( )
231.	High meat poultry dinner is suitable for infant and geriatrics.	( )
232.	Goose live sausage contain atleast 30% goose liver.	( )
233.	Chicken liver omlet contain atleast 20% chicken liver in omlet.	( )
234.	Alpine is a dry sausage of Italian brand.	( )
235.	For mechanical deboning process, larger the weight range of the carcass, lower the cost of the product.	( )
236.	Centrifugation of mechanically deboned poultry meat enhances the functional properties and storage stability.	( )
237.	In most markets, the cost of white meat considerably higher than cost of dark meat.	( )
238.	Wing is about 7.4% of the eviscerated broiler carcass.	( )
239.	For processing of goose sub-scalding is followed.	( )
240.	Springer cans are these with one end distended.	( )
241.	The pH of whole egg powder is in acidic range.	( )
242.	Technical albumen is a dehydrated product produced by centrifugation crushed egg shells.	( )
243.	Tandoori chicken is a comminuted product.	( )
244.	Chicken skin constitutes about 11-14% of whole carcass.	( )
245.	Calcium content of MDPH is higher than that of normal poultry meat.	( )
246.	Generally speaking poultry meat contains more moisture than mammalian meat.	( )
247.	Contents of an egg are enclosed in a single service container, once it is emptied, the same can never be reused.	( )
248.	Quantity of added water must be carefully controlled for preparation of chicken sausage.	( )
249.	In broilers, contribution of age is greater than weight to the meat yield.	( )
250.	In poultry processing operations the efficiency depends solely on output weight.	( )
251.	The ready to cook yield of turkeys far exceeds that of duck and broilers.	( )
252.	As defined, giblet includes neck upto 9th vertebrae, gizzard, liver and heart.	( )
253.	Meaty cuts constitute both light and dark meat.	( )

254.	Sex has no effect on composition of meat.	( )
255.	Shrinkage is inversely related to age of the bird where as directly related to holding time before slaughter.	( )
256.	In duck eviscerated yield increases with increasing age but in chicken it is not.	( )
257.	Water activity may be defined as the percentage of water in the product during preservation.	( )
258.	Rate of cooling during cool storage depends solely on size and shape of the product.	( )
259.	Freezing poultry meat by liquid nitrogen may be popularized as this is the best method of frozen storage.	( )
260.	All cured and smoked poultry can be consumed without further processing in the kitchen.	( )
261.	In curing, the efficiency of colour fixation of potassium nitrite is more than that of sodium nitrite.	( )
262.	Ducks are hard scalded as no other suitable method is available for them.	( )
263.	It is better to keep the frozen product in the refrigerator before it reaches the consumer.	( )
264.	Wet chilling is practiced because of weight gain during the process.	( )
265.	'Three point suspensions' is better than 'two point suspension' during evisceration of turkeys.	( )
266.	'Three point suspensions' is better than 'two point suspension' during evisceration of broilers .	( )
267.	Pulling of neck and windpipe is done to facilitate smooth removal of adjoining organs.	( )
268.	It is desirable to cook any product at least upto an internal temperature of 165^0F for better taste before consumption.	( )
269.	In Kosher poultry the trachea remains intact during bleeding.	( )
270.	In hard scalding, packaging of carcass is immediately necessary to prevent discolouration.	( )
271.	Stunning is done only for humanitarian ground.	( )
272.	There is no differences between dark and light meat as far as chemical composition is concerned.	( )
273.	Rough handling of birds is undesirable because of improper bleeding.	( )
274.	Overscalded carcasses are better because they require less cooking time.	( )
275.	Poorly feathered birds are subjected to more tears.	( )
276.	There will be perfect bleeding if the bird is excited.	( )

277.	Lysozyme protects the eggs from bacterial invasion by hydrolyzing of bacterial cells.	( )
278.	Carcasses from young birds have higher water to nitrogen ratio than the old ones.	( )
279.	Poultry meat is a very good source of ascorbic acid whereas it is completely absent in eggs.	( )
280.	All nutrients are uniformly distributed in different parts of egg.	( )
281.	Poultry meat is an ideal food for weight control diets because of its high nutritive value.	( )
282.	There is a linear relationship of mean weight of broiler parts with carcass.	( )
283.	The amount of water absorbed by the carcass during chilling is influenced by chilling period only.	( )
284.	The skin of broilers should be removed during processing since most of our consumers do not like it.	( )
285.	Mis-shapen eggs are produced if the birds are suffering from New Castle Disease.	( )
286.	In raw eggs avidin combines with biotin and inactivates its vitamin activity.	( )
287.	Polyphosphates are used in chilling solution for better cook yield.	( )
288.	Interaction between lysozyme and ovo-mucin is mainly responsible for maintaining the thick white in a firm gel condition.	( )
289.	Average pH of fresh egg is always nearer to 7.0.	( )
290.	Chlorinated hydrocarbons have been shown not to accumulate in the fatty portion of poultry meat.	( )
291.	Use of Co_2 for stunning of poultry speed up the rate of bleeding.	( )
292.	Complete bleeding of poultry birds is undesirable if their meat is to be stored in freezer for 6 months or more.	( )
293.	Anaesthetizing the birds make the feather pulling more difficult.	( )
294.	Freeze drying of meat results in the best quality product at a lower cost when compared with other methods of dehydration.	( )
295.	During smoking of meat, the formaldehyde is the most important bactericidal agent.	( )
296.	The type of wood used influences the flavour of smoked meat.	( )
297.	The temperature of curing solution should be kept around 30-40^{0}C during curing of poultry meat.	( )
298.	As a working figure, 1100 BTU are required to change one pound of water to vapour at common dehydration temperature.	( )

299.	Amylase is the proteolytic enzyme used in the tenderization of poultry meat.	( )
300.	Nitrite can be added in cured meat products upto 2 percent level.	( )
301.	Lower the pH of meat, higher will be its water holding capacity.	( )
302.	Faster the rate of freezing, the better is the quality of frozen meat.	( )
303.	Pickling/curing of poultry meat can be best carried out at 20-25^{0}C temperature of brine.	( )
304.	There is no effect of age of birds on the texture/tenderness of poultry meat.	( )
305.	A feed efficiency ratio of 2.50 can be attained by the efficient strains of broilers.	( )
306.	Salt is added first to the meat in emulsion type of sausage.	( )
307.	Psychrophilic bacteria cause spoilage of canned poultry meat.	( )
308.	Egg yolk contains all the fat soluble vitamins present in egg.	( )
309.	Eggs with darker yolk and fertilized eggs are more nutritious than eggs with light coloured yolks and infertile eggs.	( )
310.	Cold sterilization of poultry means destruction of micro-organism using freezing temperature.	( )
311.	Hard cooked eggs are better digested than raw eggs.	( )
312.	'Undrawn dressed poultry' means poultry that has been slaughtered for human consumption and from which the blood and feather have been removed.	( )
313.	Broilers or fryers are young chickens of either sex that are tender coated and are usually under 16 weeks of age.	( )
314.	Unlike red meats, most fat in poultry is subcutaneous rather than distributed throughout the carcass.	( )
315.	Poultry cooked under intense heat has a lower cooked shrinkage than cooked with moderate heat.	( )
316.	Broiling is a moist heat procedure of cooking used for more tender meat.	( )
317.	Cooking losses are usually higher in pressure cooked method than those for dry heat cookery.	( )
318.	Well packaged cut-up chicken stored at 20^{0}F will be in good condition for 3 months.	( )
319.	Stunning contracts the muscle which hold the feather.	( )
320.	Birds which are decapitated lose more blood than slaughtered by Kosher killing.	( )
321.	Carcasses scalded in water of 138-140^{0}F for 30 minutes are generally considered for sub scalding.	( )

322.	Anaesthetizing the birds increases the feather force by 25-50%.	( )
323.	The nitrite reacts with the haemoglobin found into form methaemoglobin.	( )
324.	Beta particles are positively charged particles from the nuclei of radio-active atoms.	( )
325.	Irradiation hastens the onset of oxidative deterioration.	( )
326.	Processing shrinkage is lower in young birds or lighter weight than in older or heavier birds.	( )
327.	The percent yield of breast of turkeys is more than that of chickens.	( )
328.	Generally the temperature of the fat is held between 200-250^{0}F during deep fat frying.	( )
329.	Lower oven temperature requires longer cooking times but materially reduce shrinkage.	( )
330.	Braising is normally used for processed meat products.	( )
331.	Poultry immersed in water heated to 160-180^{0}F for 30 to 60 seconds is considered to be slack scalded.	( )
332.	About 8.4 Mrads are required for sterilization doses for chicken meat.	( )
333.	The zoological name of common house fly is *Musca demoestica.*	( )
334.	Poultry processing plant with disposal systems which are solid wastes with water (flow away system) 4500 ltrs of water are required per 1000 birds processed and flow away system 8000 gallons.	( )
335.	Poultry industry has one of the highest accident rates in food processing industries.	( )
336.	Besides preventing struggling, relaxes the muscles by holding the feathers.	( )
337.	British recommendation for liquid egg pasteurization is 64^{0}C for $2^1/_2$ minutes.	( )
338.	Poultry is generally considered to be not suitable for canning.	( )
339.	Amount of glucose in egg white ensures solubility and stability when the powder is held in storage.	( )
340.	Preparation of powdered turkey skin, the first step is to remove water followed by removal of all fatty oils.	( )
341.	India is the second largest producer of the food.	( )
342.	Poultry term comes from the English word Poult.	( )
343.	Broilers with about 750 g dressed weight are preferred for barbecuing	( )

344.	Poultry carcass require sufficient amount of oxygen in its packaging.	( )
345.	Surimi is a poultry product.	( )
346.	EIC is the chief enforcement body for export	( )
347.	For the cold storage of eggs, relative humidity in cold storage should be 80-90%.	( )
348.	Conalbumin of egg is responsible for chelation of iron, zinc and copper.	( )
349.	Mature poultry is generally considered better for canning than younger birds.	( )

Q.3 Match the following

A.

A		B		
1.	Fishy odours in eggs	a.	Keratin	( )
2.	Cuticle	b.	*Esherichia coli*	( )
3.	BHA	c.	Quail	( )
4.	Sodium alginate	d.	Poultry slaughter	( )
5.	Filoplumes	e.	Singeing	( )
6.	Modified kosher method	f.	Anti-oxidants	( )
7.	Button	g.	Stabilizer	( )
8.	Bone darkening	h.	Chicken	( )
9.	Benzpyrene	i.	Carcinogenic	( )
10.	Ovomucoid	j.	Inhibits trypsinn	( )

B.

A		B		
1.	Haught unit	a.	Shell porosity	( )
2.	Weight loss of eggs	b.	Chelation of iron, zinc and copper	( )
3.	Dried albumen for non-food purpose	c.	Albumen quality	( )
4.	Conalbumen	d.	Candling	( )
5.	Combination with biotin and its inactivation	e.	Frozen poultry	( )
6.	Egg grading	f.	Technical albumen	( )
7.	Bone darkening	g.	Avidin	( )
8.	Curing	h.	Canned product	( )
9.	Poor bleeding of poultry	i.	Nitrosohaemochrome	( )
10.	Chicken-a-la-king	j.	Low keeping quality	( )

C.

A		B		
1.	Flavoprotein	a.	Trypsin inhibitor	( )
2.	Lysozyme	b.	Riboflavin	( )
3.	Air cell	c.	B(1-4) linkage	( )
4.	Ovomucin	d.	Pore canal	( )
5.	Thermostabilization	e.	Artherosclerosis	( )
6.	Spongy matrix	f.	Age of the egg	( )
7.	Ovomucoid	g.	Sodium lauryl sulphate	( )
8.	High density lipoprotein	h.	Bacteria culture	( )
9.	Whipping aids	i.	Coagulation of egg white	( )
10.	Bug	j.	Thick albumen	( )

D.

A		B		
1.	Hydrogen peroxide	a.	Coagulation	( )
2.	Nitrate & Nitrites	b.	Adhesive	( )
3.	Bloom gelometer	c.	Pasteurization	( )
4.	Angol cake	d.	Liquid freezing	( )
5.	Mayonaise	e.	Electrical stimulation	( )
6.	Luten	f.	Pink colour	( )
7.	Technical albumen	g.	Drum freezing	( )
8.	Calcium chloride	h.	Foaming	( )
9.	Offals	i.	Emulsifying	( )
10.	Cold shortening	j.	Yolk	( )

E.

A		B		
1.	Freezer burn	a.	Tenderness	( )
2.	Giblet	b.	Cured poultry	( )
3.	Shear press	c.	Antioxidant	( )
4.	Rigor mortis	d.	Nitrite burn	( )
5.	Pink colour	e.	Phospholipids	( )
6.	Maillard browning	f.	Improper packaging	( )
7.	Greening	g.	Desirable change	( )
8.	Stigma	h.	Whole egg powder	( )
9.	Phenols	i.	Ovulation	( )
10.	Lecithin	j.	Edible offal	( )

Key

Poultry & Egg Products Technology

Q.No. 1				
1.a	37.d	73.c	109.c	145.a
2.a	38.c	74.b	110.a	146.c
3.a	39.d	75.c	111.c	147.d
4.c	40.a	76.a	112.d	148.c
5.a	41.c	77.a	113.a	149.c
6.d	42.b	78.a	114.d	150.a
7.d	43.a	79.a	115.c	151.b
8.a	44.b	80.b	116.b	152.a
9.d	45.a	81.c	117.c	153.a
10.a	46.c	82.b	118.a	154.b
11.a	47.a	83.a	119.b	155.b
12.b	48.d	84.b	120.a	156.a
13.a	49.d	85.c	121.c	157.b
14.b	50.d	86.a	122.b	158.a
15.a	51.b	87.b	123.d	159.c
16.c	52.b	88.b	124.c	160.a
17.d	53.b	89.b	125.a	161.a
18.c	54.b	90.a	126.b	162.d
19.c	55.a	91.b	127.b	163.c
20.a	56.a	92.c	128.a	164.d
21.b	57.a	93.a	129.a	165.c
22.a	58.c	94.a	130.a	166.d
23.a	59.c	95.b	131.b	167.d
24.b	60.a	96.c	132.c	168.d
25.b	61.b	97.a	133.d	169.d
26.a	62.b	98.a	134.a	170.c
27.a	63.b	99.c	135.a	171.c
28.b	64.d	100.b	136.c	172.a
29.b	65.a	101.b	137.a	173.d
30.c	66.a	102.b	138.d	174.d
31.b	67.b	103.a	139.a	175.c
32.b	68.b	104.b	140.a	176.d
33.b	69.a	105.d	141.d	177.d
34.a	70.b	106.d	142.b	178.c
35.c	71.a	107.a	143.a	179.a
36.b	72.b	108.b	144.b	180.a

181.a				
182.d				
183.a				
184.b				
185.b				

Q.No. 2			
1.	Antibacterial proteins and peptides	21.	Maillard
2.	100	22.	Macular degeneration
3.	Riboflavin binding protein	23.	Mimosine
4.	Yolk	24.	Phosphotydal choline or sphingomyelene
5.	11	25.	Sunflower seed oil or fish oil
6.	75	26.	Mineral oil
7.	Protein hydrolysis	27.	BHT
8.	Embryo develops with its head in the small end	28.	One
9.	Due to least overlapping of bands	29.	Early in the morning o late evening
10.	Small and uniform ice crystal formation	30.	90%
11.	Elastin fibers are highly cross-linked and insoluble polymers	31.	Ice
12.	Rapid post-mortem glycolysis	32.	1-5
13.	Higher pH	33.	Different species of poultry
14.	Incapability to kill the microbes	34.	64.4°C for 2.5 minutes
15.	Higher shrinkage and microbial load	35.	Glutamic acid
16.	Gives off little heat	36.	1.9:1
17.	30-33	37.	Slower
18.	Slush ice	38.	Rancidity
19.	97	39.	Higher
20.	Reptiles	40.	Unfit

41.	5-10%	80.	Increases
42.	Improper bleeding	81.	Lysozyme
43.	2-4^{0}C for 2-4 hrs	82.	More
44.	10-12	83.	Poor
45.	10-12	84.	Coagulation
46.	0.33mm	85.	Decrease
47.	15	86.	0.09 and 0.12
48.	75-80%	87.	Salt and sugar
49.	50^{0}F	88.	Mucor
50.	0.05 % avidin	89.	Anticlock
51.	15%	90.	2
52.	0%	91.	Inert
53.	10 minutes	92.	Foaming
54.	46%	93.	Salmonella
55.	52%	94.	Leavening
56.	7.6 to 8.9	95.	Inner thick or chalaziferous layer
57.	One	96.	Heart, liver and gizzard
58.	177^{0}C for 7minutes	97.	Phenols
59.	Thicker	98.	Formaldehyde
60.	3.4%	99.	Oesophagus
61.	10%	100.	Lacithin
62.	26%	101.	Water in oil emulsion
63.	31%	102.	Flavoprotein
64.	Coagulation	103.	Triglycerides and cholesterol
65.	Yolk	104.	Livetins
66.	Albumen	105.	01
67.	Baking	106.	Biotin
68.	Ovomucoid or Ovoinhibitor	107.	Iron
69.	7.6-8.5	108.	30-60
70.	9.7	109.	15^{0}F
71.	Sugar and salt	110.	Curing ingredients
72.	Vit-C and Ca, Fe	111.	Curing adjuncts
73.	Glucosidase	112.	Immersion, stitching and pumping
74.	12^{0}C	113.	160^{0}F
75.	Iron and sulphur	114.	One
76.	Easy	115.	Antioxidants and anti-foamimg
77.	Increase	116.	Calcium mineral
78.	Gram-ve bacteria	117.	Sugar or salt
79.	Avidin	118.	Suitable humidity

119.	Increasing	158.	20250
120.	Gram-ve	159.	50
121.	Thermophilic	160.	30
122.	Increases	161.	Drying
123.	Electric shock	162.	8-10
124.	Lysozyme	163.	12-24
125.	Vitamine-B	164.	Albumen
126.	More	165.	Yolk
127.	Nutritional and managemental	166.	Sugar and alcohol
128.	Slush ice	167.	Salmonella
129.	Ear lobe	168.	Poisoning and trapping
130.	Hard	169.	90 volts for 1-2
131.	More	170.	160^{0}F
132.	Difficult	171.	2-4
133.	Vitamin-C	172.	25-31^{0}F
134.	Drip	173.	Polyethylene and cryovac
135.	3 pieces	174.	Water activity
136.	'B' quality	175.	5-7
137.	USDA 'C' quality	176.	Instant coffee and meat
138.	Thigh and drumstick	177.	100
139.	20% in raw and 25%	178.	Myoglobin
140.	Humerus	179.	62.5^{0}C for 3.5
141.	0.01%	180.	Myonnaise
142.	14%	181.	$1^{1}/_{2}$ hrs
143.	2%	182.	Lysozyme
144.	15%	183.	*Clostridium botulium* PA363679
145.	45%, 5% and 50%	184.	5 to 1-2
146.	Barbecued chicken moist heat cooked	185.	125-130 ^{0}F
147.	95%	186.	138-140^{0}F
148.	140^{0}F for $3^{1}/_{2}$	187.	*S. cerevisae* and Glucose oxidase
149.	54%	188.	½ to $^{1}/_{3}$
150.	60	189.	Salt and sugar
151.	54%	190.	Ovomucin
152.	15^{0}C	191.	65% &88%
153.	Increases	192.	400^{0}F
154.	Restrict	193.	$20^{3}/_{4}$
155.	Increases	194.	Calcium mineral and ascorbic acids
156.	1300-1500FPM	195.	5.2%
157.	36000	196.	Ovomucoid

197.	10-11^{0}C	217.	Glucose protein reaction
198.	Gossypol	218.	4.4^{0}C
199.	Sulphhydril	219.	Below 2.0
200.	Phenol	220.	Welded stainless steel bar
201.	400^{0}F	221.	Glucose-Cepahlin
202.	160-170^{0}F	222.	*S. senftenburg* 775 W
203.	-10^{0}F	223.	97
204.	31 to 25^{0}F	224.	Reptiles
205.	0^{0}F	225.	Mallard
206.	Modified Kosher	226.	Macular degeneration
207.	123-130^{0}F for 90-120sec	227.	Mimosine
208.	40^{0}F	228.	Phosphotydal choline or sphingomyelene
209.	155^{0}F	229.	Sunflower seed oil or fish oil
210.	Pyrethrum or allethrin	230.	Mineral oil
211.	*Blatta orientalis*	231.	BHT
212.	Increased	232.	One
213.	Filoplumes	233.	Early in the morning o late evening
214.	250^{0}F for 65	234.	90
215.	Sulphhydral	235.	Ice
216.	Hard scalding		

Q.No. 3																	
1.	T	41.	T	81.	T	121.	T	161.	F	201.	F	241.	T	281.	F	321.	F
2.	F	42.	T	82.	F	122.	F	162.	T	202.	T	242.	F	282.	F	322.	T
3.	F	43.	F	83.	F	123.	T	163.	T	203.	F	243.	T	283.	T	323.	F
4.	T	44.	T	84.	T	124.	T	164.	F	204.	F	244.	F	284.	F	324.	F
5.	F	45.	T	85.	T	125.	F	165.	T	205.	F	245.	T	285.	F	325.	F
6.	F	46.	F	86.	F	126.	T	166.	F	206.	T	246.	T	286.	T	326.	T
7.	T	47.	F	87.	F	127.	F	167.	F	207.	F	247.	F	287.	T	327.	T
8.	F	48.	F	88.	T	128.	T	168.	T	208.	F	248.	T	288.	F	328.	T
9.	T	49.	T	89.	F	129.	F	169.	F	209.	F	249.	T	289.	T	329.	F
10.	F	50.	T	90.	F	130.	T	170.	T	210.	F	250.	F	290.	T	330.	T
11.	T	51.	T	91.	T	131.	T	171.	T	211.	T	251.	F	291.	F	331.	F
12.	F	52.	F	92.	T	132.	T	172.	F	212.	F	252.	T	292.	T	332.	F
13.	T	53.	F	93.	T	133.	T	173.	F	213.	F	253.	F	293.	F	333.	F
14.	F	54.	T	94.	F	134.	T	174.	F	214.	T	254.	T	294.	F	334.	T
15.	F	55.	F	95.	T	135.	T	175.	F	215.	F	255.	F	295.	F	335.	F
16.	F	56.	T	96.	F	136.	F	176.	F	216.	F	256.	T	296.	T	336.	T
17.	T	57.	T	97.	T	137.	T	177.	T	217.	T	257.	F	297.	T	337.	T
18.	T	58.	T	98.	F	138.	T	178.	T	218.	F	258.	F	298.	F	338.	T
19.	T	59.	T	99.	F	139.	F	179.	T	219.	T	259.	F	299.	T	339.	F
20.	T	60.	T	100.	T	140.	F	180.	F	220.	F	260.	F	300.	F	340.	T
21.	T	61.	T	101.	F	141.	T	181.	T	221.	F	261.	F	301.	F	341.	T
22.	F	62.	T	102.	T	142.	F	182.	F	222.	T	262.	F	302.	F	342.	T
23.	F	63.	T	103.	T	143.	T	183.	T	223.	F	263.	T	303.	T	343.	F
24.	F	64.	T	104.	F	144.	F	184.	F	224.	T	264.	T	304.	F	344.	T
25.	F	65.	T	105.	F	145.	F	185.	F	225.	T	265.	F	305.	F	345.	F
26.	F	66.	F	106.	F	146.	F	186.	T	226.	T	266.	T	306.	T	346.	F
27.	F	67.	F	107.	T	147.	T	187.	T	227.	T	267.	F	307.	T	347.	F
28.	T	68.	T	108.	F	148.	T	188.	T	228.	T	268.	T	308.	F	348.	T
29.	F	69.	F	109.	T	149.	F	189.	F	229.	T	269.	F	309.	T	349.	T
30.	T	70.	F	110.	F	150.	F	190.	T	230.	T	270.	T	310.	F	350.	T
31.	F	71.	F	111.	F	151.	T	191.	F	231.	T	271.	T	311.	F		
32.	T	72.	F	112.	F	152.	T	192.	T	232.	T	272.	F	312.	T		
33.	F	73.	T	113.	F	153.	T	193.	T	233.	T	273.	F	313.	T		
34.	T	74.	F	114.	F	154.	F	194.	F	234.	T	274.	F	314.	T		
35.	T	75.	F	115.	F	155.	F	195.	T	235.	T	275.	F	315.	T		
36.	F	76.	F	116.	F	156.	F	196.	F	236.	T	276.	T	316.	F		
37.	T	77.	F	117.	T	157.	T	197.	T	237.	T	277.	F	317.	F		
38.	F	78.	F	118.	F	158.	T	198.	F	238.	T	278.	F	318.	T		
39.	F	79.	T	119.	T	159.	F	199.	T	239.	T	279.	T	319.	T		
40.	T	80.	T	120.	T	160.	F	200.	F	240.	T	280.	F	320.	F		

Q.No. 4									
A		B		C		D		E	
1.	b	1.	c	1.	b	1.	c	1.	f
2.	a	2.	a	2.	c	2.	f	2.	j
3.	f	3.	f	3.	f	3.	a	3.	a
4.	g	4.	b	4.	j	4.	h	4.	g
5.	e	5.	g	5.	i	5.	i	5.	b
6.	d	6.	d	6.	d	6.	j	6.	h
7.	c	7.	e	7.	a	7.	b	7.	d
8.	h	8.	i	8.	e	8.	d	8.	i
9.	i	9.	j	9.	g	9.	g	9.	c
10.	j	10.	h	10.	h	10.	e	10.	e

8

Food Biotechnology

Q.1 Multiple type questions

1. ______________________ is commonly known as baker's or brewer's yeast:
   a. *Saccharomyces cerevisiae* b. *Lactobacillus lactis*
   c. *Yersinia pestis* d. None of the above
2. Example of Food acidulant is
   a. Lactic acid b. Gluconic acid
   c. Citric acid d. All of the above
3. Example of flavour and fragrances used in food industry is
   a. Yeast extract b. Glutamic acid
   c. Nucleic acid hydrolysates d. All of the above
4. The most well known use of fermentation in food industry is the batch production of
   a. Alcoholic beverages b. Meat products
   c. Milk products d. None of the above
5. Which one is a aerotolerant bacteria
   a. *Corynebacterium glutamicum* b. Lactic acid bacteria
   c. Aspergillus niger d. None of the above
6. *Xanthomonas compestris* is used to produce
   a. Acaia gum b. Babool gum
   c. Xanthan gum d. None of the above
7. Ultrafiltration can be used to separate macromolecules such as
   a. Enzymes from liquid b. Unwanted proteins from liquid
   c. Both of the above d. None of the above

8. ________________bacteria usually dominates the LAB in naturally fermented sausages.
   a. Lactobacilli
   b. Lactococci
   c. Micrococci
   d. None of the above
9. Under anaerobic conditions nitrates and nitrites serve as
   a. Oxidizing agent
   b. Reducing agent
   c. All of the above
   d. None of the above
10. Nisin is a example of
   a. Toxin
   b. Bacteriocins
   c. Bacteriophase
   d. None of the above
11. Technique that permits the manipulation of genes of one organisms which can subsequently be introduced into genome of another organisms of same or other species.
   a. Mutation
   b. Transplantation
   c. Transgenesis
   d. None of the above
12. Intramuscular fat is commonly known as
   a. Marbling
   b. Seem fat
   c. Lean fat
   d. All of the above
13. Poultry fat is mostly
   a. Intramuscular
   b. Intermuscular
   c. Subcutaneous
   d. None of the above
14. Beef and Buffalo meat can be differentiated on the basis of
   a. Meat colour
   b. Fat colour
   c. Carotene content
   d. All of the above
15. Cow and its progeny slaughter is permissible in
   a. West Bengal
   b. Kerala
   c. Uttar Pradesh
   d. Both a and b
16. Name the scientist who discovered penicillin
   a. Alexander Fleming
   b. John Lister
   c. Spallanzani
   d. None of the above

17. Industrial microbiology began in 1857 with the invention of

    a. Alexander Fleming b. Spallanzani
    c. Louis Pasteur d. None of the above

18. ____________is a device in which a substrate of low value is utilized by living cells or enzymes to generate a product of higher value

    a. Bioreactor b. Fermenters
    c. Both of the above d. None of the above

19. The process in which information gathered from small fermenters when used to predict proper fermentation condition for very large scale

    a. Scale up process b. Upgradation
    c. Modernization d. None of the above

20. The device used to introduce air into the fermenter broth is called

    a. Impellar b. Baffles
    c. Sparger d. None of the above

21. Antifoaming agents used in fermenter is

    a. Sunflower oil b. Olive oil
    c. Silicons d. All of the above

22. Industrial microbiology began in 1857 with the invention of

    a. Alexander Fleming b. Spallanzani
    c. Louis Pasteur d. None of the above

23. Which is not antimicrobial compound

    a. Bacteriocins b. Lysozyme
    c. Streptomycin d. Gum Acacia

24. Name the texture foaming agent

    a. Xanthan gum b. Olive oil
    c. Silicons d. All of the above

25. Which one is not a fermented product of commercial benefit

    a. Cheese b. Paneer
    c. Yoghurt d. Soy sauce

26. Name the bacteria used as probiotic
    a. Lactobacillus
    b. Lactococci
    c. Streptococcus
    d. All of the above
27. Culture used to start a good fermentation is known as
    a. Mother culture
    b. Starter culture
    c. Finisher culture
    d. None of the above
28. The technique used to express desirable genes by genetic modification is called
    a. Hybridization
    b. Recombinant DNA technology
    c. Mutagenesis
    d. None of the above
29. Successful application of genetic modification for food bioprocessing requires development and use of food grade
    a. Vector
    b. Plasmid
    c. Both of the above
    d. None of the above
30. Plasmids do not contain antibiotics resistance genes as marker but consist DNA sequence from microorganisms and recognized as
    a. Generally Recognised as safe (GRAS)
    b. Generally Recognised as unsafe (GRAUS)
    c. No relation to safety
    d. None of the above
31. National Programme on Organic Production is administered by
    a. Ministry of agriculture
    b. Ministry of Food Processing
    c. Ministry of Commerce
    d. None of the above
32. Which is true for organic production?
    a. Grass fed animals
    b. Use of synthetic chemicals
    c. Use of growth hormones
    d. Use of antibiotics
33. Name the standard dealing with organic food internationally
    a. IFOAM basic standards
    b. EU regulation No. 1804/1999
    c. ALINORM 99/22A
    d. All of the above

34. Pigs expressing an *E.coli* salivary phytase produced ____ phosphorus manure

    a. Low
    b. High
    c. No effect
    d. None of the above

35. Transgenic chickens expressing bacterial beta-galactosidase hydrolyze lactose in intestine and use ________________ as energy source

    a. Fat
    b. Protein
    c. Sugar
    d. None of the above

36. *Saccharomyces cerevisiae*, commonly known as______________, was first eukaryote to have its genome sequenced in 1996.

    a. Baker's yeast
    b. Brewer's yeast
    c. Both of the above
    d. None of the above

37. Biological catalysts used to facilitate and speed up metabolic reactions in living organisms are called

    a. Hormones
    b. Enzymes
    c. Both of the above
    d. None of the above

38. Name the organisms in which DNA coding of chymosin was done which is used in curdling of milk in cheese manufacturing

    a. *Kluyveromyces lactis*
    b. *Escherichia coli*
    c. *Aspergillus niger* var. awamori
    d. All of the above

39. Scaled up culture system is called

    a. Bioreacter
    b. Fermentation
    c. Culturing
    d. None of the above

40. Typical example of aerotolerant is

    a. Lactic acid bacteria
    b. *Aeromonas hydrophila*
    c. *E. coli*
    d. None of the above

41. Prokaryotes used for commercial production of xanthan gum is

    a. Kluyveromyces
    b. *Xanthomonas campestris*
    c. *Escherichia coli*
    d. One of the above

42. Name the organism used for commercial production of glutamate, an amino acid that is an important flavor enhancer of food

    a. Kluyveromyces
    b. *Xanthomonas campestris*
    c. *Corynebacterium glutamicum*
    d. None of the above

43. Indicate the example of food acidulant
    a. Lactic acid
    b. Gluconic acid
    c. Citric acid
    d. All of the above
44. Yeast starter cultures used to produce powdery surface or fruity or alcoholic aroma is
    a. *Debaryomyces hansenii*
    b. *Candida famata*
    c. Both of the above
    d. All of the above
45. Soppressata molisana is a
    a. Traditional dry sausage of Italy
    b. Traditional semidry sausage of Italy
    c. Traditional fermented sausage of Italy
    d. None of the above
46. The most important mechanism to ensure safety of fermented meat is the development of
    a. Lactic acid
    b. Acetic acid
    c. Bacteriocins
    d. All of the above
47. Meat adulteration is punishable under
    a. PFA act
    b. AGMARK act
    c. BIS act
    d. None of the above
48. Ammonical odour is the characteristics of
    a. Chevon
    b. Mutton
    c. Veal
    d. Venison
49. Urine like odour is the characteristics of
    a. Chevon
    b. Mutton
    c. Pork
    d. Venison
50. Indicate the meat in which fat is absent in muscles and confined to the body cavity
    a. Rabbit meat
    b. Mutton
    c. Veal
    d. Venison

51. Highest value of glycogen (2.08%), Refractive index (53.5) and iodine value (71-86) is in

    a. Buffalo meat
    b. Beef
    c. Horse meat
    d. Venison

52. Indicate the PCR technique which is capable to identify animal species of spoiled meat

    a. RFLP
    b. Multiplex PCR
    c. RAPD
    d. Nested PCR

53. Indicate the correct combination of genes responsible for tenderness in different animals

    a. CLPG in sheep
    b. Myostatin in beef
    c. RN in pork
    d. All of the above are correct

54. ____________ is a device in which a low value substrate is utilized by living cells or enzymes to generate a high value product.

    a. Bioreactors
    b. Fermenters
    c. Both of the above
    d. None of the above

55. Advantages of purified enzymes over cell mass are

    a. Catalyses on specific reactions
    b. Perform the reaction at optimal condition
    c. No interference of other substrate
    d. All of the above are correct

Q.2. Fill in the blanks

1. ______________ is a process of bioconversion of organic substances by micro-organisms and/or enzymes.
2. A culture used to start a food fermentation is known as a ______________.
3. ______________________ are biological catalysts used to facilitate and speed up metabolic reactions in living organisms.
4. GMO's stands for ________________________.
5. _________________ is a key enzyme used in manufacture of cheeses.
6. A scaled up culture system is called ______________ system.
7. ________________________ bacteria is able to grow at pH below 5.0.
8. ________________________ is an undefined carbon source for cell culture development.

9. High degree of purified amino acids can be produced by ________ filteration technique.
10. ____________________ starter cultures are used to improve the competitiveness of the starter strains and to prevent the growth of food pathogens .
11. PFA stands for ____.
12. AFST(I) stands for____.
13. MFPO stands for ____.
14. Nitrate and Nitrite are commonly used compounds in curing but these compounds are also responsible for the formation of ________________ which is carcinogenic in nature.
15. PCR stands for ________________________________.
16. Foaming in fermenter is due to ________________________________.
17. When a batch of culture is subsequently fed with fresh nutrient medium without removing the growing microbial culture is called ____________________________.
18. The process of mutants production by exposure of microbial strains to mutagenic chemicals or UVrays to induce changes in their genome is called ____________________________.
19. ___________________is a natural process of transfer of genetic material among closely related microbial species by physical relation between donor and recipient microorganism.
20. Serological tests of meat speciation are mainly based on________________ reaction.
21. Agar Gel Immunodiffusion test for meat speciation was first time performed by ____________________________.
22. PCR-RFLP technique utilizes PCR alongwith the digestion with specific___.
23. Forensically Informative Nucleotide Sequencing utilizes the sequencing with ______________________.
24. Genetically modified animals are those in which specific changes are introduced by genetic engineering using ________________________.
25. First time pigs were engineered to produce omega-3 fatty acids through the expression of ______________________________.
26. ________________________ is GM yeast used in brewing and baking applications.

27. Microarrays can also be referred as ______________________.
28. Bacteria are sensitive to acidic pH but LAB can grow at pH______________.
29. The major application of immobilized cell to food biotechnology is in the production of ______________________.
30. The most well known use of fermentation is in food industry is in batch production of ______________________ and process is strictly anaerobic.
31. Proteolysis may help in flavor development by releasing ____________.
32. The growth of metabolism of lactic acid bacteria is affected by____________.
33. The temperature of slow fermentation is ________________ in which nitrate is reduced to nitrite by bacterial nitrate reductase activity.
34. The temperature of fast fermentation is ________________ depending on sausage diameter.
35. During cured ham production,________________ are responsible for free amino acid generation and flavor development.
36. ________________ is the most important group of starter culture.
37. ____________________techniques is based on the combination of generic enzymatic and fluorescence resonance energy transfer methods of meat speciation.
38. Cow milk containing specific human proteins which may be helpful in treating the human emphysema can be proced by ______________________.
39. 'Healthy pork' produced by transgenic technique denotes the pig carrying plant gene and posses higher amount of ____________________ in muscles of pigs.
40. PSE in pigs is____________________ genetically governed character.

Q.3 State whether the following statements are 'True' or 'False'

1.	Safety of fermented food products is enhanced through reduction of toxic compounds.	( )
2.	Aseptic does not mean sterile.	( )
3.	Micro-organisms are sensitive to changes in pH.	( )
4.	LAB represents the most important group of starter organisms.	( )
5.	In long ripened meat products significant amount of bigenic amines are produced.	( )
6.	a_w plays an important role in growth of micro-organisms.	( )
7.	Probiotics are harmful organism in food for human health.	( )

8.	Besides nutrients, soy also contributes some antinutritional factors.	( )
9.	Marbling is a characteristic feature of buffalo meat.	( )
10.	LAB has antagonistic effect against Salmonella.	( )
11.	Fermentaion is the process of bioconversion of organic substances by organisms and enzymes.	( )
12.	In dot blot technique of meat speciation, sample antigen binds with nitrocellulose or yanogens bromide activated nitrocellulose membrane with specific antibody.	( )
13.	ELISA can identify meat adulteration upto 2%.	( )
14.	PAGE is mainly based on the protein differentiation.	( )
15.	In DNA hybridization technique, DNA or cloned DNA is hybridized with target DNA and detected by colour development or audiography.	( )
16.	Mitochondrial cytochrome b gene is most commonly used in PCR technique of meat speciation.	( )
17.	GMO based feed is permissible in organic livestock farming.	( )
18.	Carcass development in organically reared animals is more.	( )
19.	GM foods came into market first time in 1990s and were mainly transgenic plant products.	( )
20.	Genetically modified pigs are available who can absorb plant phosphorus more efficiently.	( )
21.	Chymosin produced by recombinant microorganisms is commercially utilized for cheese manufacture.	( )
22.	Genetic based diagnostic and identification of great importance for specificity, sensitivity and speedy testing.	( )
23.	Fermentation can enhance the nutritional value of food through biosynthesis of health beneficial components.	( )
24.	Aerotolerant organisms are those they do not use oxygen but can tolerate its presence.	( )
25.	Fungi are much more tolerant to acidic conditions than bacteria.	( )
26.	Enzymes are usually easier to immobilize than cells.	( )
27.	Typical example of enzyme and cell immobilization in bioprocess is conversion of starch into lactic acid by *Lactobacillus casei.*	( )
28.	Most commercially important fermentation processes are aerobic.	( )
29.	Final stage of fermentation cycle is cell lysis.	( )
30.	The first food area likely to be impacted by recombinant DNA is the production of enzyme.	( )
31.	Spray dried milk does not affect the lysine availability while roller dried milk reduced the availability.	( )

32.	Beef and poultry irradiated at -40^{0}C shows only minor changes in protein and is free from free radicals and antithiamine properties.	( )
33.	About 50% thiamine lost in fish meat when treated with 0.6 Mrad while niacin is unaffected.	( )
34.	Fermentation of meat products reduces their shelf life.	( )
35.	Fermented sausages are usually made from comminuted meat.	( )
36.	Italian salamis are manufactured from pork only.	( )
37.	Lactobacilli dominate the lactic acid bacteria in naturally fermented sausages.	( )
38.	All genera of LAB are capable to produce bacteriocins.	( )
39.	Recent PCR techniques are capable to identify the spoiled meat	( )
40.	In Random Amplified Polymorphic DNA fingerprinting (RAPD) fingerprints are generated to identify the meat species.	( )

Q.4 Match the following statements

A.

	A		B	
1.	Veal	a	Cell culture production	( )
2.	Venison	b	Urea	( )
3.	Urine like odour	c	Calf meat	( )
4.	White meat	d	Whey concentrates	( )
5.	GM yeast	e	Deer meat	( )
6.	Biogenic amines	f	Poultry meat	( )
7.	ORBIT	g	Pork	( )
8.	Ultrafiltration	h	*Saccharomyces cerevisiae*	( )
9.	Nitrogen source	i	Fermentation	( )
10.	Batch process	j	Beef	( )

B.

	A		B	
1.	HAL gene	a	Extreme muscling	( )
2.	RN gene	b	Myofibrillar protein breakdown	( )
3.	Callipyge gene	c	Postmortem proteolysis	( )
4.	REM gene	d	Muscularity	( )
5.	Myostatin	e	PSE meat	( )
6.	Calpain gene	f	Regulate homeostasis	( )

7.	BT-47 CAST system gene	g	High muscle glycogen	( )
8.	Melanocortin Receptor gene	h	Tenderization	( )
9.	IGF2 gene	i	Large cross sectional area	( )
10.	Spinach gene	j	Negative growth factor	( )

C.

A		B		
1.	RYR1 in pork	a	Ultrafiltartion	( )
2.	Pituitary derived somatotropin Hormone	b	Single cell protein supplement in animal feed	( )
3.	Chymosin	c	Probiotics	( )
4.	Commercial galactosidase	d	Rate of rigor mortis	( )
5.	Cheese whey utilization	e	*Saccharomyces cerevisiae*	( )
6.	Yeast mass	f	Homologies or difference at species and sub-species level	( )
7.	Lactobacillus	g	Salted and sundried meat	( )
8.	GM yeast	h	Growth and carcass composition	( )
9.	Ribotyping	i	Genetically engineered *E.coli*	( )
10.	Kaddid	j	*Kluyveromyces lactis*	( )

Food Biotechnology Key

Q.No. 1		Q.No. 2		Q.No. 3	
1.a	41.b	1.	Biotransformation	1.	T
2.c	42.c	2.	Starter cuture	2.	T
3.d	43.d	3.	Enzymes	3.	T
4.a	44.c	4.	Genetically Modified Organisms	4.	T
5.b	45.a	5.	Rennet	5.	T
6.c	46.d	6.	Bioreactor	6.	T
7.c	47.a	7.	Acid loving microbes/acidophiles	7.	F
8.a	48.b	8.	Yeast extract	8.	T
9.c	49.c	9.	Membrane	9.	F
10.b	50.a	10.	LAB	10.	T
11.c	51.c	11.	Prevention of Food Adultration Act	11.	T
12.a	52.d	12.	Association of Food Scientists & Technologists (India)	12.	T
13.c	53.d	13.	Meat Food Products Order	13.	T
14.d	54.c	14.	Nitrosamine	14.	T
15.d	55.d	15.	Polymerase Chain Reaction	15.	T
16.a		16.	Protein and microbial components	16.	T
17.c		17.	Fed batch culture	17.	F
18.c		18.	Mutagenesis	18.	F
19.a		19.	Conjugation	19.	T
20.c		20.	Antigen and antibody	20.	T
21.d		21.	Oucherlony in 1948	21.	T
22.c		22.	Enzyme	22.	T
23.d		23.	Phylogenetic analysis	23.	T
24.a		24.	Cisgenesis or Transgenesis	24.	T
25.b		25.	Roundworm gene	25.	T
26.a		26.	Saccharomyces cerevisiae	26.	T
27.b		27.	Biochip or DNA chips	27.	T
28.b		28.	<5.0	28.	T
29.c		29.	Enzymes	29.	T
30.a		30.	Alcoholic beverages	30.	T
31.c		31.	Amines and amino acids	31.	T
32.a		32.	Oxygen	32.	T
33.d		33.	15-16^{0}C	33.	T
34.a		34.	22-25^{0}C for 2 to 3 days	34.	F
35.c		35.	Muscle aminopeptides	35.	T
36.c		36.	Lactic acid bacteria	36.	T
37.b		37.	DNA invader assay	37.	T

Q.No. 1		Q.No. 2		Q.No. 3	
38.d		38.	Transgenosis	38.	T
39.a		39.	Unsaturated fatty acids	39.	T
40.a		40.	50%	40.	T

Q.No. 4

A		B		C	
1.	c	1.	e	1.	d
2.	e	2.	g	2.	h
3.	g	3.	a	3.	i
4.	f	4.	i	4.	j
5.	h	5.	j	5.	a
6.	i	6.	b	6.	b
7.	j	7.	c	7.	c
8.	d	8.	f	8.	e
9.	b	9.	d	9.	f
10.	a	10.	h	10.	g

9

Wool Technology

Q.1 Multiple type questions

1. Wooled sheep skin is called
   a. Pelt
   b. Fur
   c. Hair
   d. None of the above
2. Carbonizing refers to
   a. Chemical treatment of wool
   b. Washing with alkali
   c. Washing with detergent
   d. None of the above
3. Average weight of bale (wool pack) is
   a. 500-600Kg
   b. 300-400 Kg
   c. 100-200 Kg
   d. None of the above
4. Term 'burr'is used for
   a. Vegetable matter
   b. Twigs
   c. Straw
   d. All of the above
5. Indian wool is basically of
   a. Fine wool
   b. Carpet wool
   c. Clothing wool
   d. None of the above
6. Crimp is
   a. Natural waviness in wool fibre
   b. Developed waviness in wool fibre
   c. Straightness of wool
   d. None of the above
7. Total length of wool fiber in natural condition is called
   a. Staple length
   b. Fiber length
   c. Fleece length
   d. None of the above

8. The entire coat sheared from sheep is called

    a. Fleece
    b. Felt
    c. Drip
    d. None of the above

9. Find out the correct match of 'Hank' (length of textile material)

    a. 560 Yards for wool
    b. 840Yards for cotton and silk
    c. 360 Yards for linen
    d. All of the above

10. Yolk term is used in wool technology for

    a. Grease
    b. Suint
    c. Grease+ Suint
    d. None of the above

11. The annual total yield of greasy wool per sheep in India is

    a. 1-1.5 Kg
    b. 2-2.5 Kg
    c. 4-5 Kg
    d. 3-4 Kg

12. Wool insoluble protein is called

    a. Keratin
    b. Collagen
    c. Elastin
    d. None of the above

13. Ideal colour of wool is

    a. White or creamy white
    b. Brown
    c. Red
    d. Reddish brown

14. Average weight of bale (wool pack) is

    a. 500-600kg
    b. 300-400 kg
    c. 100-200 kg
    d. None of the above

15. Indicate the name of fur producing animals

    a. Angora rabbit
    b. Mink
    c. Otters
    d. All of the above

16. Camel hairs are

    a. One layer structure
    b. Two layered structure
    c. Three layered structure
    d. None of the above

17. Pashmina is produced from

    a. Goat
    b. Sheep
    c. Rabbit
    d. All of the above

18. Name the breeds of goats producing Pashmina
    a. Chegu and Changathangi
    b. Swarlla
    c. Ranabati
    d. All of the above
19. The cuticle pattern in very fine wool is
    a. Coronal
    b. Coronal reticulate
    c. Reticulate
    d. None of the above
20. Fiber diameter of fine wool is
    a. 17-20 mμ
    b. 22-24 mμ
    c. 25-32 mμ
    d. None of the above
21. Kemp wool is mainly found in the body of sheep at
    a. Back
    b. Flank
    c. Shoulder
    d. Britch
22. ISI grading of wool is based on
    a. Blood system
    b. Micron value
    c. Wool quality
    d. None of the above
23. American system of wool grading is based on
    a. Blood system
    b. Micron value
    c. Wool quality
    d. None of the above
24. Wool fiber has______degree of resilience
    a. Low
    b. High
    c. Medium
    d. None of the above
25. The refractive index of wool is
    a. 1.551-1.555
    b. 1.321-1.451
    c. 1.666-1.700
    d. None of the above
26. Which one is true for wool fiber
    a. Acid resistance
    b. Alkali sensitive
    c. Water sensitive
    d. All of the above
27. Chemical shearing is obtained by using
    a. Cyclohexamide
    b. Cyclophosphamide
    c. Hexmetaphosphate
    d. None of the above

28. Wool oiling is generally performed with
    a. Vegetable oil
    b. Mineral oil
    c. Animal fat
    d. None of the above
29. Separation of wool fibers cling together is called
    a. Willowing
    b. Spinning
    c. Carding
    d. None of the above
30. Dyeing of wool is generally performed by
    a. Acid dyes
    b. Alkali dyes
    c. Both of the above
    d. None of the above
31. Most suitable dye for cellulose fiber is
    a. Acid dyes
    b. Alkali dyes
    c. Azoic dyes
    d. None of the above
32. Core sampling technique is generally preferred for the measurement of
    a. Fiber diameter
    b. Grease content
    c. Vegetable matter
    d. All of the above
33. Shrinkage in raw wool can be determined by
    a. Core test
    b. Medulla test
    c. Bale test
    d. None of the above
34. Wool purity can be assessed by
    a. Acid test
    b. Alkali test
    c. Benzene test
    d. None of the above
35. Wool ash can be determined by
    a. Acid test
    b. Alkali test
    c. Benzene test
    d. ASTM-D test
36. Fiber diameter can be measured by
    a. Air flow instrument
    b. Projection Microscope
    c. Both of the above
    d. None of the above
37. Wool fineness can be measured by
    a. Crimpometer
    b. Ermascope
    c. Air flow apparatus
    d. All of the above

38. Fiber length can be assessed by
    a. AL-2000 method
    b. OFDA-4000 method
    c. WIRA fiber diagram machine
    d. All of the above
39. Fiber damage can be identified by
    a. Acid test
    b. Alkali test
    c. Methylene blue absorption test
    d. None of the above
40. Wool pH can be determined by
    a. Calorimetric method
    b. Indicator solution method
    c. Electronic pH meter
    d. All of the above

Q.2 Fill in the blanks

1. Opening and dusting of wool is generally performed by ____________.
2. The number of fibres present in per unit area of sheep body is called ____________________.
3. Any type of wool used for preparation of cloths except carpet and pulled wool is called ______________________________.
4. Vegetable matter like twings, straw etc present in wool is referred as ___________________.
5. ______________ is a wool pack containing a specified minimum weight of wool.
6. Annual production of wool from a given flock, state or country is called____.
7. __________________ is a numerical expression which defines the degree of fineness.
8. Total wool obtained from a single sheep in one sheering is referred as_____.
9. Grading of wool is mainly based upon_____________________________ ____________.
10. 560 yards of worsted yarn is known as ______________________.
11. ______________________________are the natural impurities of the wool.
12. Most important dye used in wool is ____________________________.
13. Sampling of wool is generally performed by_________________________.
14. Finest wool may be obtained from the ___________________________ part of sheep.

15. Projection microscope method is generally used to measure the__________.
16. The specific gravity of pure non- modulated wool fibre in benzene is __________________.
17. ______________________is the total length of wool fibre after removing the crimps or waviness by straightening.
18. ________________is a slot of wool produced from one flock.
19. ________________is the ability of wool to reflect light.
20. Ability of fiber against pulling, stretching and breakage is termed as_______.
21. Lanolin is purified wool wax and can be utilized as _______________.
22. Fineness of wool is directly related to the ____________________________.
23. Wool is ______________________ conductor of heat.
24. Hairs contain medulla and developed from_____________________ follicles.
25. Wooled sheep skin is called_____________________________.
26. Fur contains _______________________shaped medulla.
27. _________________ temperature is essential for Pashmina production.
28. Wool fiber is originated from follicles found in ____________________layer of skin.
29. Cuticle pattern of Merino wool is_______________________________.
30. Honeycomb like structure filled with air cell and sometimes like hollow tube is called______________________.

Q.No.3. State whether the following statements are 'True' or 'False'

1.	Scouring refers to cleaning of wool by washing with acid.	( )
2.	Wool burns slowly and becomes porous and brittle on heating.	( )
3.	Wool is neither soluble in cold nor hot water.	( )
4.	Wool is comparatively sensitive to alkalis than acids.	( )
5.	Wool is a very good insulator of electricity.	( )
6.	Refractive index of wool is 1.553 to 5.	( )
7.	Wool is elastic and hygroscopic in nature.	( )
8.	In true wool medulla is absent.	( )
9.	Wool is generally produced by secondary Follicles.	( )
10.	Wool containing small parts of shrubs is called 'Twigs'.	( )

11.	If the burrs are not removed they cause considerable difficulty in all the manufacturing processes.	( )
12.	Damage done to wool fiber during processing is tested by colorimetric tests.	( )
13.	Crimpness is expressed in terms of crimp per inch or centimeters.	( )
14.	Fur refers to any animal skin or part thereof with hair, fleece or fur fibres.	( )
15.	Lanolin is a purified wool greese.	( )
16.	The total number of fibres in course wool is greater than Merino wool.	( )
17.	Air flow method is a common method for fibre diameter measurement.	( )
18.	Greater crimp in wool indicates course wool.	( )
19.	Twist denotes turns per inch in textile.	( )
20.	Scouring is the method of cleaning wool with alkali and detergents.	( )
21.	Alkali weakens the wool adversely.	( )
22.	Crimpiness is the ability of wool to provide softness to wool.	( )
23.	In normal wool, moisture varies between 12-17 percent.	( )
24.	Projection microscope is used to know the medullation percentage.	( )
25.	Clean wool obtained after scouring is called tinged wool.	( )
26.	The number of fibres present per unit area of sheep's body is called fleece density.	( )
27.	Bradford system is considered as best for classification of wool fibres based on fiber diameter.	( )
28.	Scouring is the method of cleaning wool with alkali and detergents.	( )
29.	Wool is the fibers present over sheep skin.	( )
30.	Wool consist of 33% keratin, 26% dust, 28% suint, 12% fat and 1% minerals.	( )

Q.4 Match the following

A.

A		B		
1.	Bating	a.	Hollow centre of hair	( )
2.	Hank	b.	Short fibres	( )
3.	Hogget wool	c.	Dried sweat secretion of sheep	( )